Advanced Nanoscale Materials for Thermoelectric Applications

Advanced Nanoscale Materials for Thermoelectric Applications

Editors

Ting Zhang
Peng Jiang

Basel • Beijing • Wuhan • Barcelona • Belgrade • Novi Sad • Cluj • Manchester

Editors

Ting Zhang
Institute of Engineering Thermophysics
Chinese Academy of Sciences
Beijing
China

Peng Jiang
Dalian Institute of Chemical Physics
Chinese Academy of Sciences
Dalian
China

Editorial Office
MDPI
St. Alban-Anlage 66
4052 Basel, Switzerland

This is a reprint of articles from the Special Issue published online in the open access journal *Nanomaterials* (ISSN 2079-4991) (available at: www.mdpi.com/journal/nanomaterials/special_issues/nanoscale_materials_thermoelectric_applications).

For citation purposes, cite each article independently as indicated on the article page online and as indicated below:

Lastname, A.A.; Lastname, B.B. Article Title. *Journal Name* **Year**, *Volume Number*, Page Range.

ISBN 978-3-7258-0012-4 (Hbk)
ISBN 978-3-7258-0011-7 (PDF)
doi.org/10.3390/books978-3-7258-0011-7

© 2024 by the authors. Articles in this book are Open Access and distributed under the Creative Commons Attribution (CC BY) license. The book as a whole is distributed by MDPI under the terms and conditions of the Creative Commons Attribution-NonCommercial-NoDerivs (CC BY-NC-ND) license.

Contents

About the Editors vii

Preface ix

Ting Zhang
Advanced Nanoscale Materials for Thermoelectric Applications
Reprinted from: *Nanomaterials* **2023**, *13*, 3165, doi:10.3390/nano13243165 1

Cordelia Zimmerer, Frank Simon, Sascha Putzke, Astrid Drechsler, Andreas Janke and Beate Krause
N-Type Coating of Single-Walled Carbon Nanotubes by Polydopamine-Mediated Nickel Metallization
Reprinted from: *Nanomaterials* **2023**, *13*, 2813, doi:10.3390/nano13202813 4

Lijun Zhao, Haiwei Han, Zhengping Lu, Jian Yang, Xinmeng Wu, Bangzhi Ge, et al.
Realizing the Ultralow Lattice Thermal Conductivity of Cu_3SbSe_4 Compound via Sulfur Alloying Effect
Reprinted from: *Nanomaterials* **2023**, *13*, 2730, doi:10.3390/nano13192730 20

Tristan da Câmara Santa Clara Gomes, Nicolas Marchal, Flavio Abreu Araujo and Luc Piraux
Flexible Active Peltier Coolers Based on Interconnected Magnetic Nanowire Networks
Reprinted from: *Nanomaterials* **2023**, *13*, 1735, doi:10.3390/nano13111735 33

Numan Salah, Neazar Baghdadi, Shittu Abdullahi, Ahmed Alshahrie and Kunihito Koumoto
Thermoelectric Power Generation of TiS_2/Organic Hybrid Superlattices Below Room Temperature
Reprinted from: *Nanomaterials* **2023**, *13*, 781, doi:10.3390/nano13040781 48

Min Sun, Yu Liu, Dongdan Chen and Qi Qian
Multifunctional Cu-Se Alloy Core Fibers and Micro–Nano Tapers
Reprinted from: *Nanomaterials* **2023**, *13*, 773, doi:10.3390/nano13040773 62

Min Sun, Pengyu Zhang, Guowu Tang, Dongdan Chen, Qi Qian and Zhongmin Yang
High-Performance n-Type Bi_2Te_3 Thermoelectric Fibers with Oriented Crystal Nanosheets
Reprinted from: *Nanomaterials* **2023**, *13*, 326, doi:10.3390/nano13020326 68

Javier Gainza, Federico Serrano-Sánchez, Oscar J. Dura, Norbert M. Nemes, Jose Luis Martínez, María Teresa Fernández-Díaz, et al.
Reduced Thermal Conductivity in Nanostructured $AgSbTe_2$ Thermoelectric Material, Obtained by Arc-Melting
Reprinted from: *Nanomaterials* **2022**, *12*, 3910, doi:10.3390/nano12213910 76

Ruiyi Li, Xiao Yang, Jian Li, Ding Liu, Lixin Zhang, Haisheng Chen, et al.
Pre-Ball-Milled Boron Nitride for the Preparation of Boron Nitride/Polyetherimide Nanocomposite Film with Enhanced Breakdown Strength and Mechanical Properties for Thermal Management
Reprinted from: *Nanomaterials* **2022**, *12*, 3473, doi:10.3390/nano12193473 88

M. Almasoudi, Numan Salah, Ahmed Alshahrie, Abdu Saeed, Mutabe Aljaghtham, M. Sh. Zoromba, et al.
High Thermoelectric Power Generation by SWCNT/PPy Core Shell Nanocomposites
Reprinted from: *Nanomaterials* **2022**, *12*, 2582, doi:10.3390/nano12152582 99

Sedong Kim
Study on the Characteristics of the Dispersion and Conductivity of Surfactants for the
Nanofluids
Reprinted from: *Nanomaterials* **2022**, *12*, 1537, doi:10.3390/nano12091537 . . . . **114**

About the Editors

Ting Zhang

Ting Zhang is currently a Professor at the Institute of Engineering Thermophysics (IET), Chinese Academy of Sciences (CAS), and the Director of the Centre for Advanced Technology at Nanjing Institute of Future Energy System (China). He obtained his Ph.D. degree in Condensed Matter Physics from Beijing Normal University (China) in 2014 and his B.E. degree in Materials Science and Engineering from Xi'an University of Technology (China) in 2009. He was an Assistant Professor at the Institute of Electrical Engineering, CAS (2014-2015) and a Research Fellow at Nanyang Technological University in Singapore (2015-2019). Professor Ting Zhang specializes in the areas of thermoelectrics, hydrogen storage, thermal transport and recovery, and fiber-based devices for energy harvesting and storage.

Peng Jiang

Peng Jiang is currently a Professor at the Dalian Institute of Chemical Physics (DICP), Chinese Academy of Sciences (CAS). He obtained his B.E. degree (1998-2002) from Shandong Universityand Ph.D. degree (2002-2007) from the Institute of Physics (IOP), CAS. After completing his post doc (2007-2011) at the Lawrence Berkeley National Laboratory, he joined DICP in 2011. His research focuses on thermoelectric materials and devices, photothermoelectric detectors, and surface science.

Preface

Thermoelectric modules can achieve energy conversion between heat and electricity, which are generally used for power generation or electronic refrigeration, and are important in solving energy crises and environmental pollution. However, the efficiency of existing thermoelectric materials is inferior to that of heat engines under the same operating conditions. Nanomaterials such as superlattices, quantum dots, nanowires, and nanocomposites are considered some of the most effective methods for decoupling thermoelectric parameters (e.g., electric conductivity, thermal conductivity, and Seebeck coefficient) to enhance the performance of thermoelectric materials. Within this framework, this reprint is a collection of research articles authored by expert authorities, focusing on nanoscale materials for thermoelectric applications.

Ting Zhang and Peng Jiang
Editors

Editorial

Advanced Nanoscale Materials for Thermoelectric Applications

Ting Zhang [1,2,3,4,5]

1. Nanjing Institute of Future Energy System, Nanjing 211135, China; zhangting@iet.cn
2. Institute of Engineering Thermophysics, Chinese Academy of Sciences, Beijing 100190, China
3. University of Chinese Academy of Sciences, Beijing 100049, China
4. Innovation Academy for Light-Duty Gas Turbine, Chinese Academy of Sciences, Beijing 100190, China
5. University of Chinese Academy of Sciences, Nanjing 211135, China

Recently, there has been growing academic interest in researching thermoelectric materials that exhibit energy conversion capability between thermal energy and electricity, providing solutions to energy crises and environmental pollution [1–6]. Generally, the efficiency of existing thermoelectric materials still has potential for improvement compared with traditional heat engines under the same operating conditions [7–9]. Nanomaterials, such as superlattices, quantum dots, nanowires, and nanocomposites, are considered one of the most effective materials for decoupling thermoelectric parameters, thus enhancing the performance of thermoelectric materials [10–14]. Although some relevant research has already been published, there is still great potential to further investigate the preparation, measurements, devices, and applications associated with thermoelectric nanoscale materials.

This Special Issue intends to summarize the advanced developments towards highly efficient thermoelectric nanomaterials and applications. In this Special Issue, we present nine high-quality original papers from the field of advanced nanoscale materials, with contributions from more than 50 authors worldwide.

In terms of inorganic thermoelectric materials, nanomaterials have an exceptional performance due to their narrow band gap, high electrical conductivity and low thermal conductivity. Zhao et al. found that a copper-based chalcogenide Cu_3SbSe_4 could achieve a maximum ZT value of 0.72 at 673 K due to a sulfur alloying effect, which widened the band gap, increased the effective carrier mass, and scattered phonons [Contribution 1]. Additionally, the use of nanowire networks is considered an effective method for manufacturing thermoelectric modules. Tristan et al. demonstrated a flexible thermoelectric module by embedding Co-Fe nanowires in a polymer film with a power factor of 4.7 mW/mK^2 [Contribution 2]. The fabricated thermocouple operated as a Peltier cooler and achieved an equivalent cooling of 1.2 mW. Furthermore, thermoelectric fibers are promising for wearable applications due to their flexibility. Sun et al. successfully prepared flexible Cu-Se alloy core fibers using a thermal drawing method and obtained a high power factor of 1.2 mW/mK^2, higher than that for bulk polycrystals [Contribution 3]. The Cu-Se fiber was applied to thermal–electric response with 5% measurement uncertainty. Using the same method, n-type Bi_2Te_3 fibers were prepared and the microstructure during the annealing process was explored [Contribution 4]. The reported Bi_2Te_3 fiber demonstrated an enhanced ZT value of 1.05 at room temperature after the Bridgman annealing processes.

Carbon nanotubes (CNTs) are widely used in the fields of electronics, energy and functional materials. The novel preparation technique makes it practical to develop high-thermoelectric-performance CNTs. Zimmerer et al. explored an environmentally friendly technique to coat single-walled carbon nanotubes (SWCNTs) with nickel using polydopamine (PDA) as an adhesion promoter [Contribution 5]. The results show that the SWCNTs modified by PDA have good dispersion and a homogeneous coating. The Seebeck coefficient of the obtained SWCNTs was reversed from positive to negative and reached −19 uV/K for the n-type application. Moreover, Almasoudi et al. polymerized CNTs with polypyrrole (PPy) in

Citation: Zhang, T. Advanced Nanoscale Materials for Thermoelectric Applications. *Nanomaterials* **2023**, *13*, 3165. https://doi.org/10.3390/nano13243165

Received: 30 November 2023
Accepted: 13 December 2023
Published: 18 December 2023

Copyright: © 2023 by the author. Licensee MDPI, Basel, Switzerland. This article is an open access article distributed under the terms and conditions of the Creative Commons Attribution (CC BY) license (https://creativecommons.org/licenses/by/4.0/).

situ to form one-dimensional core–shell nanocomposites [Contribution 6]. The thermoelectric properties, including power factor and ZT value, were optimized to 0.36 mW/mK^2 and 0.09, respectively. Using the prepared sample, a thermoelectric generator that can generate a maximum power of 24 nW at a temperature difference of 40 K was designed.

Hybrid thermoelectric materials combine the flexibility of organic materials with the high performance of inorganic materials and serve to further enhance the applications of flexible thermoelectric materials. For example, Li et al. produced boron nitride/polyetherimide (PEI) composite films via a casting–hot pressing method [Contribution 7]. The tensile strength of the composite film reached 102.7 MPa giving it a potential application in a flexible circuit substrate. Furthermore, Salah et al. developed a TiS$_2$/organic hybrid superlattice (TOS), which had an optimized power factor of 0.1 mW/mK^2 at a temperature of 233 K [Contribution 8]. Further studies suggest that TOS devices have better application prospects in cool environments than those at room temperature. In addition, Kim et al. prepared and studied several cellulose nanocrystal (CNC) aqueous solutions with surfactant aqueous solutions [Contribution 9]. As a result, the non-covalent dispersion method showed effective dispersion and stability.

In conclusion, this Special Issue, entitled "Advanced Nanoscale Materials for Thermoelectric Applications", collates state-of-the-art achievements in the relevant fields of research, including CNTs, thermoelectric fibers and high-performance films. This Special Issue provides broad insights into the valuable research in these rapidly advancing and interdisciplinary fields.

Funding: This work was supported by the National Key Research and Development Program of China (2023YFB3809800), the National Natural Science Foundation of China (52172249), the Chinese Academy of Sciences Talents Program (E2290701), and the Special Fund Project of Carbon Peaking Carbon Neutrality Science and Technology Innovation of Jiangsu Province (BE2022011).

Acknowledgments: We greatly acknowledge the support and contributions of the Special Issue authors and reviewers.

Conflicts of Interest: The author declares no conflict of interest.

List of Contributions:

1. Zhao, L.; Han, H.; Lu, Z.; Yang, J.; Wu, X.; Ge, B.; Yu, L.; Shi, Z.; Karami, A.M.; Dong, S.; Hussain, S.; Qiao, G.; Xu, J. Realizing the Ultralow Lattice Thermal Conductivity of Cu$_3$SbSe$_4$ Compound via Sulfur Alloying Effect. *Nanomaterials* **2023**, *13*, 2730.
2. Da Câmara Santa Clara Gomes, T.; Marchal, N.; Abreu Araujo, F.; Piraux, L. Flexible Active Peltier Coolers Based on Interconnected Magnetic Nanowire Networks. *Nanomaterials* **2023**, *13*, 1735.
3. Sun, M.; Liu, Y.; Chen, D.; Qian, Q. Multifunctional Cu-Se Alloy Core Fibers and Micro-Nano Tapers. *Nanomaterials* **2023**, *13*, 773.
4. Sun, M.; Zhang, P.; Tang, G.; Chen, D.; Qian, Q.; Yang, Z. High-Performance n-Type Bi$_2$Te$_3$ Thermoelectric Fibers with Oriented Crystal Nanosheets. *Nanomaterials* **2023**, *13*, 326.
5. Zimmerer, C.; Simon, F.; Putzke, S.; Drechsler, A.; Janke, A.; Krause, B. N-Type Coating of Single-Walled Carbon Nanotubes by Polydopamine-Mediated Nickel Metallization. *Nanomaterials* **2023**, *13*, 2813.
6. Almasoudi, M.; Salah, N.; Alshahrie, A.; Saeed, A.; Aljaghtham, M.; Zoromba, M.S.; Abdel-Aziz, M.H.; Koumoto, K. High Thermoelectric Power Generation by SWCNT/PPy Core Shell Nanocomposites. *Nanomaterials* **2022**, *12*, 2582.
7. Li, R.; Yang, X.; Li, J.; Liu, D.; Zhang, L.; Chen, H.; Zheng, X.; Zhang, T. Pre-Ball-Milled Boron Nitride for the Preparation of Boron Nitride/Polyetherimide Nanocomposite Film with Enhanced Breakdown Strength and Mechanical Properties for Thermal Management. *Nanomaterials* **2022**, *12*, 3473.
8. Salah, N.; Baghdadi, N.; Abdullahi, S.; Alshahrie, A.; Koumoto, K. Thermoelectric Power Generation of TiS$_2$/Organic Hybrid Superlattices Below Room Temperature. *Nanomaterials* **2023**, *13*, 781.
9. Kim, S. Study on the Characteristics of the Dispersion and Conductivity of Surfactants for the Nanofluids. *Nanomaterials* **2022**, *12*, 1537.

References

1. Liu, H.; Fu, H.; Sun, L.; Lee, C.; Yeatman, E.M. Hybrid energy harvesting technology: From materials, structural design, system integration to applications. *Renew. Sustain. Energy Rev.* **2021**, *137*, 110473. [CrossRef]
2. Shi, X.L.; Zou, J.; Chen, Z.G. Advanced Thermoelectric Design: From Materials and Structures to Devices. *Chem. Rev.* **2020**, *120*, 7399. [CrossRef]
3. Zhou, C.; Lee, Y.K.; Yu, Y.; Byun, S.; Luo, Z.-Z.; Lee, H.; Ge, B.; Lee, Y.-L.; Chen, X.; Lee, J.Y.; et al. Polycrystalline SnSe with a thermoelectric figure of merit greater than the single crystal. *Nat. Mater.* **2021**, *20*, 1378. [CrossRef] [PubMed]
4. Gao, M.; Wang, P.; Jiang, L.; Wang, B.; Yao, Y.; Liu, S.; Chu, D.; Cheng, W.; Lu, Y. Power generation for wearable systems. *Energy Environ. Sci.* **2021**, *14*, 2114. [CrossRef]
5. Tang, X.; Li, Z.; Liu, W.; Zhang, Q.; Uher, C. A comprehensive review on Bi_2Te_3-based thin films: Thermoelectrics and beyond. *Interdiscip. Mater.* **2022**, *1*, 88. [CrossRef]
6. Shen, Y.; Han, X.; Zhang, P.; Chen, X.; Yang, X.; Liu, D.; Yang, X.; Zheng, X.; Chen, H.; Zhang, K.; et al. Review on Fiber-Based Thermoelectrics: Materials, Devices, and Textiles. *Adv. Fiber Mater.* **2023**, *5*, 1105. [CrossRef]
7. Shen, Y.; Wang, Z.; Wang, Z.; Wang, J.; Yang, X.; Zheng, X.; Chen, H.; Li, K.; Wei, L.; Zhang, T. Thermally drawn multifunctional fibers: Toward the next generation of information technology. *InfoMat* **2022**, *4*, e12318. [CrossRef]
8. Masoumi, S.; O'Shaughnessy, S.; Pakdel, A. Organic-based flexible thermoelectric generators: From materials to devices. *Nano Energy* **2022**, *92*, 106774. [CrossRef]
9. Abid, N.; Khan, M.; Shujait, S.; Chaudhary, K.; Ikram, M.; Imran, M.; Haider, J.; Khan, M.; Khan, Q.; Maqbool, M. Synthesis of nanomaterials using various top-down and bottom-up approaches, influencing factors, advantages, and disadvantages: A review. *Adv. Colloid Interface Sci.* **2022**, *300*, 102597. [CrossRef] [PubMed]
10. Mao, J.; Liu, Z.; Zhou, J.; Zhu, H.; Zhang, Q.; Chen, G.; Ren, Z. Advances in thermoelectrics. *Adv. Phys.* **2018**, *67*, 69. [CrossRef]
11. Yang, L.; Chen, Z.-G.; Dargusch, M.S.; Zou, J. High Performance Thermoelectric Materials: Progress and Their Applications. *Adv. Energy Mater.* **2018**, *8*, 1701797. [CrossRef]
12. Wu, Y.; Finefrock, S.W.; Yang, H.R. Nanostructured thermoelectric: Opportunities and challenges. *Nano Energy* **2012**, *1*, 651. [CrossRef]
13. Liu, Z.; Hong, T.; Xu, L.; Wang, S.; Gao, X.; Chang, C.; Ding, X.; Xiao, Y.; Zhao, L. Lattice expansion enables interstitial doping to achieve a high average ZT in n-type PbS. *Interdiscip. Mater.* **2023**, *2*, 161. [CrossRef]
14. Su, L.; Wang, D.; Wang, S.; Qin, B.; Wang, Y.; Qin, Y.; Jin, Y.; Chang, C.; Zhao, L. High thermoelectric performance realized through manipulating layered phonon-electron decoupling. *Science* **2022**, *375*, 1385–1389. [CrossRef] [PubMed]

Disclaimer/Publisher's Note: The statements, opinions and data contained in all publications are solely those of the individual author(s) and contributor(s) and not of MDPI and/or the editor(s). MDPI and/or the editor(s) disclaim responsibility for any injury to people or property resulting from any ideas, methods, instructions or products referred to in the content.

Article

N-Type Coating of Single-Walled Carbon Nanotubes by Polydopamine-Mediated Nickel Metallization

Cordelia Zimmerer [1], Frank Simon [2], Sascha Putzke [1], Astrid Drechsler [2], Andreas Janke [1] and Beate Krause [3,*]

[1] Institute of Polymer Materials, Leibniz-Institut für Polymerforschung Dresden e.V. (IPF), Hohe Str. 6, 01069 Dresden, Germany; zimmerer@ipfdd.de (C.Z.); putzke@ipfdd.de (S.P.)
[2] Institute of Physical Chemistry and Polymer Physics, Leibniz-Institut für Polymerforschung Dresden e.V. (IPF), Hohe Str. 6, 01069 Dresden, Germany; frsimon@ipfdd.de (F.S.); drechsler@ipfdd.de (A.D.)
[3] Institute of Macromolecular Chemistry, Leibniz-Institut für Polymerforschung Dresden e.V. (IPF), Hohe Str. 6, 01069 Dresden, Germany
* Correspondence: krause-beate@ipfdd.de; Tel.: +49-351-4658-736

Abstract: Single-walled carbon nanotubes (SWCNTs) have unique thermal and electrical properties. Coating them with a thin metal layer can provide promising materials for many applications. This study presents a bio-inspired, environmentally friendly technique for CNT metallization using polydopamine (PDA) as an adhesion promoter, followed by electroless plating with nickel. To improve the dispersion in the aqueous reaction baths, part of the SWCNTs was oxidized prior to PDA coating. The SWCNTs were studied before and after PDA deposition and metallization by scanning and transmission electron microscopy, scanning force microscopy, and X-ray photoelectron spectroscopy. These methods verified the successful coating and revealed that the distribution of PDA and nickel was significantly improved by the prior oxidation step. Thermoelectric characterization showed that the PDA layer acted as a p-dopant, increasing the Seebeck coefficient S of the SWCNTs. The subsequent metallization decreased S, but no negative S-values were reached. Both coatings affected the volume conductivity and the power factor, too. Thus, electroless metallization of oxidized and PDA-coated SWCNTs is a suitable method to create a homogeneous metal layer and to adjust their conduction type, but more work is necessary to optimize the thermoelectric properties.

Keywords: thermoelectric; carbon nanotubes; polydopamine; nickel

1. Introduction

Applying carbon nanotubes (CNTs) as fillers in metal or other material composites is promising to create materials with unique performance. Metallized CNTs can be considered steady organic metals without the need for further activation by doping or charge transfer to achieve powerful electron carrier mobility [1]. Especially for lightweight metal-matrix composites [2], electrical and thermal conductive adhesives [3], impact protection and vibration damping [4], metallized carbon allotropes have great potential as compatibilized fillers [5–7], for sensors [8–10], catalysis [8,11], and energy storage [12].

CNTs have exceptional thermal and electrical properties. They are able to generate thermoelectric voltage when a temperature gradient is applied [13] and can thus be used for thermoelectric (TE) applications [14–19]. With regard to the thermoelectric performance of CNTs, however, it has to be noted that it is lower than that of traditional TE materials, such as half-Heusler compounds, clathrates, silicides, antimonides, and tellurides [13,20].

The thermoelectric (TE) properties are characterized by the Seebeck coefficient S. This parameter S is calculated from the generated thermoelectric voltage (U) divided by the applied temperature difference (ΔT) (Equation (1)). p-conductive electrical behaviour (conduction based on the directional movement of defect electrons (holes)) is indicated by a positive S-value and n-conductive behaviour (conduction through freely movable electrons) by a negative S-value. The second parameter used is the power factor PF, which is calculated from the product of the squared Seebeck coefficient S and the volume conductivity σ (Equation (2)) [21,22].

$$S = \frac{U}{\Delta T} \tag{1}$$

$$PF = S^2 \cdot \sigma \tag{2}$$

Nonoguchi et al. [15] have conducted a screening experiment with a wide range of additives, which include phosphine-containing and imine-containing molecules as dopants for single-walled CNTs (SWCNTs). Different polymeric dopants were described by Piao et al. [14] for the modification of pieces of SWCNT buckypaper by immersing them overnight in the respective solutions. In both papers, it was reported that the additives can change the initial positive S-value of SWCNTs to other positive S-values but also to negative values. Hata et al. [18] showed that the Seebeck coefficient could be reduced from 62.3 µV/K for pure CNTs to values between −30.1 and −44.1 µV/K when nanotubes were wrapped with surfactants. Tzounis et al. [16] described polyetherimide (PEI)-based composites with SWCNTs, whereby the Seebeck coefficient of the pure SWCNTs at 31 µV/K could be increased up to 55 µV/K for a composite containing 4.4 vol% SWCNTs. Mytafides et al. [17] prepared n-type SWCNT films by solution mixing of SWCNTs with cetyltrimethylammonium bromide (CTAB). All these doping methods have in common that the p-conductive properties of the SWCNTs are enhanced or changed to n-conductive by bringing polymers to their surface in the solution.

The preparation of CNT composites faces, however, serious challenges:
- Creating well-dispersed systems in the matrix material;
- ensuring high interfacial adhesion to the matrix material;
- avoiding damage, such as structure defects and shortening, to the CNTs and thus altering their properties during composite preparation, especially for metal matrices.

To fulfil the first two requirements, highly oxidative, toxic, and, in general, environmentally harmful chemicals [23–25] or expensive, energy-consuming processes have been widely applied [23–25], and result in chemical modifications of the surface to improve the wetting of the CNTs and, therefore, the compatibility with the partner material(s) of the composite. A further disadvantage of chemical surface functionalization is the difficult control of the balance between functionalization and the damage degree of CNTs.

New paths open up with bioinspired concepts in material science. A versatile adhesion promotion concept from the field of synthetic biology has been successfully transferred to the technical sphere in recent years, adapting the mussel adhesive derivative dopamine (DA) as an environmentally friendly bridge between different types of materials [26–28]. DA can be applied in a "green" process as a water-based coating. On solid surfaces, it auto-oxidizes spontaneously under oxygen exposure and forms a thin film of oligomers and polymers, the so-called polydopamine (PDA), with high intra- and intermolecular interaction potential [29,30]. The universal adhesion mechanism is explained by the various functional groups of DA and the fact that loosely adhering layers are replaced by DA during its polymerization directly on the surface. This reaction is accompanied by a volume contraction, which leads to interlocking with the surface [28]. In particular, the size of the CNTs and their rod shape geometrically favour stable functionalization with PDA as a closed wrapping. The polymerisation of PDA leads to a covalently cross-linked, enveloping film around the tubular structure, and the shrinkage during polymerisation leads to close contact with the CNT carbon backbone. Furthermore, the high polarity of the hydroxyl groups on the surfaces of PDA leads to a significant change in the electronic properties of the CNTs due to their interaction with π-electrons in the benzene rings of the CNTs [31–33]. Studies show that CNTs are capable of absorbing various metal ions in solutions [34,35]. Molecular dynamics simulations have shown that hydroxyl and carboxyl functional groups on the CNT surface lead to more effective adsorption of Cu (II) ions [36].

In general, the contact resistance of the CNTs is affected by any functionalization. Yet, PDA films are only a few molecular layers thick and exhibit redox-active behaviour that enables electron transfer. With PDA coating, high compatibility of the CNTs with and adhesion to metals are expected. PDA functionalization of CNTs was applied for decoration with gold nanoparticles [37] and in electrodeposition composite coating formation [38,39]. Elsewhere, electroless metallization with silver and copper based on PDA as an adhesion promoter was demonstrated on tungsten carbide microparticles and on alumina nanoparticles [40].

Different metallization methods [1,23,24,41,42] such as powder metallurgy techniques, electroless plating, electroplating, physical and chemical vapour deposition, flame-, arc-, and supersonic spraying, make a colourful bouquet for different metals or applications [5]. The most well-known method is autocatalytic electroless plating. It has high mass production importance because it is scalable and straightforward in a non-vacuum environment. Metals such as those given in Table 1 can be processed from aqueous solutions under suitable conditions (reducing agent, temperature, bath composition, pH, and corresponding catalytic surface) [11]. Especially Ni possesses technical relevance because the plated layers are almost not porous, smooth and hard, and show more uniform thickness and better corrosion resistance. Against the background of the suitability of the metals as thermoelectric materials, the Seebeck coefficients S of the pure metals are given in Table 1. Low, mostly single-digit values with positive and negative signs are observed for the metals in Table 1. It is interesting to note that nickel (Ni) and cobalt (Co) show particularly high negative S-values with −19 and −20 µV/K, respectively. No S-values are given for the metal alloys, as they vary for different compositions.

Table 1. Selected electrolessly depositable metals, their Seebeck coefficients, and metal alloys for main industrial applications.

Metals	Reference	Seebeck Coefficient S [µV/K] @300 K	Metal Alloys	Reference
Ni	[5,10,11,23,43–47]	−19 [48]	CuNi, NiCo, PdNiP, NiWP,	[11,43]
Cu	[11,43,49–51]	1.7 [48]	CuNi, CuCo, CuAu, CuCd,	[11]
Co	[11]	−1.7 [52]	CuCo, NiCo, CuCd, CuAu, PdCoP	[11]
Cd	[11]	2.6 [13]	CuCd,	[11]
Ag, Ag-NP *	[9,11,43,53]	1.5 [13]	AgAu,	[11]
Au	[11,43]	1.9 [13]	AuSn, AgAu, CuAu, AuIn	[11,43]
Pt	[11]	−4.9 [48]		
Pd	[11]	−10.7 [13]	PdCoP, PdZnP, PdNiP,	[11]
Rh	[11]	1 [54]		
Cr	[11]	12 [54]		
Zn	[11]	2.4 [13]	ZnCo, NiZn, PdZnP	[11]
Sn	[11,43]	−1 [55]	AuSn, SnPb	[11]
Co	[11,43]	−20 [54]	CuCo, PdCoP,	[11]

* nanoparticle (NP).

The aim of the present study is to develop a technique for coating SWCNTs with nickel using PDA as an adhesive coupler to investigate how the coating affects their thermoelectric properties (Figure 1). If possible, the Seebeck coefficients of the SWCNTs should be reversed to generate n-type SWCNTs. To the best of our current information, PDA-functionalized and electroless nickel-coated SWCNTs have not been fabricated yet as hybrid nanoparticles. Both the modification of the SWCNTs with PDA and the subsequent nickel deposition take place in an aqueous medium. Unmodified SWCNTs are difficult to disperse in water due to their polarity. Therefore, oxidized SWCNTs were investigated as a parallel approach in the modification steps. They are better dispersible and separable in water due to their significantly more polar character. The research study will investigate whether the two surface modifications lead to higher electroless metal deposition rates when the SWCNTs are more strongly separated at the beginning of the synthesis. The influence of all modification steps on their thermoelectric properties will be investigated.

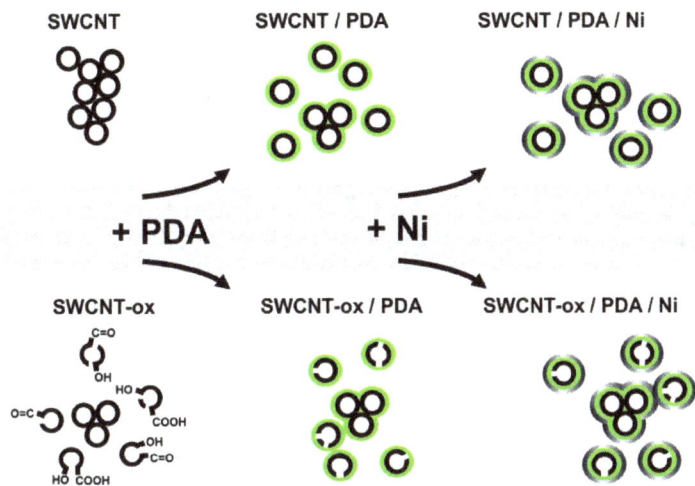

Figure 1. Schematic drawing of the SWCNTs; process steps to functionalize and metallize yielding in structure modification.

2. Materials and Methods

2.1. Materials

Single-walled carbon nanotubes (SWCNTs) of the type Tuball™ grade 75% (OCSiAl Europe S.à r.l., Leudelingen, Luxembourg) with diameters less than 2 nm and lengths larger than 1 µm were used as electrically conductive filler. Structural details are described in [56]. Tuball™ SWCNTs were selected because earlier investigations showed that this material has a very high Seebeck coefficient and is suitable for changing the thermoelectric conduction type [57,58].

Nitric acid (AnalaR NORMAPUR Nitric Acid, 65%, VWR International GmbH, Darmstadt, Germany) was used to oxidize the SWCNTs, and dilution deionized water (ELGA PURELAB Plus, Veolia Water Technologies Deutschland GmbH, Celle, Germany) was applied.

2.2. SWCNT Treatment

2.2.1. SWCNT Oxidation

Three grams of SWCNTs was added to 150 mL of HNO_3 (AnalaR NORMAPUR Nitric Acid, 65%, VWR International GmbH, Darmstadt, Germany) and dispersed in an ultrasonic bath USC600TH (VWR International GmbH, Darmstadt, Germany) at 120 W for 10 min. The reaction mixture was boiled for 2 h under reflux, then cooled with ice water and diluted with 100 mL deionized water via the reflux condenser. The SWCNTs were separated by suction filtration using a PTFE filter with a 1.2 µm pore size (Sartorius Stedim Biotech, Göttingen, Germany) and washed neutrally. Before further use, the SWCNTs were dried for 8 h at 120 °C in a vacuum oven. The oxidized SWCNTs appear in the paper as "SWCNT-ox".

2.2.2. Dispersion and PDA Deposition

For TRIS buffer preparation, 10 mmol/L 2-amino-2-(hydroxymethyl)propane-1,3-diol (99.9%, Roche Diagnostics GmbH, Grenzach-Wyhlen, Germany) was dissolved in deionized water (0.055 µS/cm) and adjusted to pH = 8.5 using hydrochloric acid (37%, VWR International GmbH, Darmstadt, Germany).

Briefly, 0.3 g of SWCNT or SWCNT-ox was dispersed in 750 mL of TRIS buffer. An ultrasonic processor (UP400St, Hielscher Ultrasonics GmbH, Teltow, Germany) with a sonotrode H3 (Hielscher Ultrasonics GmbH, Teltow, Germany) was used for vigorous mixing at an amplitude of 60% with a cycle of 1 for 5 min.

The SWCNT and SWCNT-ox dispersions were diluted with an additional 1.25 L of TRIS buffer. Then, 4 g of dopamine hydrochloride solution (2 g/L, 99%, Thermo Fisher GmbH, Bremen, Germany) (DA) was added. The suspensions were stirred at 250 rpm for 60 min to allow DA oxidative polymerization to endow the nanotube surfaces with PDA. Afterwards, they were suction filtered using PTFE filters with a 1.2 µm pore size and washed twice with 100 mL of deionized water.

2.2.3. Metallization of the PDA-SWCNTs

An electroless nickel plating bath series from MacDermit Enthone, Waterbury, CT, USA, was used to metallize the CNTs. The SWCNT/PDA and SWCNT-ox/PDA sample dispersions were first immersed in a colloidal palladium activator bath (UDIQUE 879W) for 3 min at 30 °C under stirring, then filtered off by suction filtration using PTFE filters and washed two times with deionized water (2× 100 mL). Next, CNTs were immersed in an accelerator bath solution (UDIQUE 8810) for 2.5 min at 50 °C with stirring, then separated from the solution by suction filtration with PTFE filters. In the metal coating bath process, the SWCNT/PDA and SWCNT-ox/PDA were metallized using a nickel-plating bath (UDIQUE 891, UDIQUE 892, UDIQUE 893) for 8 min at 30 °C. The pH of the nickel bath was adjusted to pH 9 with ammonium hydroxide (28–30 wt% solution of ammonia in water: $NH_3 \cdot H_2O$, Acros Organics B.V.B.A., Geel, Belgium). After metallization, the CNTs, now named SWCNT/PDA/Ni and SWCNT-ox/PDA/Ni, were separated from the solution by suction filtration with PTFE filters, washed twice with deionized water (2× 100 mL), and dried under vacuum at 25 °C for 30 min.

2.3. Characterization

For the transmission electron microscopy (TEM) study, a drop of the freshly prepared aqueous dispersion was placed on a carbon-coated TEM grid and dried in air. The TEM images were collected with a Libra120 (Carl Zeiss GmbH, Oberkochen, Germany).

For scanning electron microscopy (SEM), the dry powder was put on the grid. A Carl Zeiss Ultra plus SEM with an SE2 detector at 3 kV (Carl Zeiss Microscopy Deutschland GmbH, Oberkochen, Germany) was used for these studies. Before imaging, the surfaces were coated with 3 nm platinum.

All XPS studies were carried out by means of an Axis Ultra photoelectron spectrometer (Kratos Analytical, Manchester, UK). The spectrometer was equipped with a monochromatic Al Kα (h·ν = 1486.6 eV) X-ray source of 300 W at 15 kV. The kinetic energy of the photoelectrons was determined with a hemispheric analyzer set to pass energy of 160 eV for wide-scan spectra and 20 eV for high-resolution spectra. With Scotch double-sided adhesive tape (3M Company, Maplewood, MN, USA), the powdery SWCNT samples were prepared as thick films on a sample holder, enabling the samples' transport to the recipient of the XPS spectrometer. During all measurements, electrostatic charging of the sample was avoided by means of a low-energy electron source working in combination with a magnetic immersion lens. Later, all recorded peaks were shifted by the same value that was necessary to set the component peak Gr showing the sp^2-hybridized carbon atoms of the graphite-like lattice ($-^{Gr}\underline{C}=^{Gr}\underline{C}- \leftrightarrow =^{Gr}\underline{C}-^{Gr}\underline{C}=$) to 283.99 eV [59]. Quantitative elemental compositions were determined from peak areas using experimentally determined sensitivity factors and the spectrometer transmission function. Spectrum background was subtracted, according to Shirley [60]. The high-resolution spectra were deconvoluted by means of the Kratos spectra deconvolution software (Vision Processing, version 2.2.9 [2011], provided by Kratos

Analytical, Manchester, UK). Free parameters of component peaks were their binding energy (BE), height, full width at half maximum, and the Gaussian–Lorentzian ratio.

For the AFM investigation, freshly cleaved HOPG (highly oriented pyrolytic graphite) was dipped into CNT-water suspensions for 5 s. The remaining drops were removed by a paper tissue. The AFM measurements were done in tapping mode using a Dimension FastScan AFM (Bruker-Nano, Billerica, MA, USA) and silicon nitride sensors with a sharpened silicon tip (ScanAsyst-Fluid+) (Bruker-Nano, Billerica, MA, USA). They have a nominal spring constant of 0.7 N/m and a nominal resonance frequency of 150 kHz, and the tip radius is 2 nm. Height images (surface morphology) and phase images were taken simultaneously. According to Magonov [61], the scan conditions were chosen either to get stiffness contrast (free amplitude > 100 nm, setpoint amplitude ratio 0.8) or adhesion contrast (free amplitude < 20 nm, setpoint amplitude ratio 0.7) in the phase image.

The thermoelectric (TE) characterization was carried out using a Seebeck measuring device developed at IPF Dresden. More details are given in [57,62]. The measurements were performed at 40 °C with temperature differences between the two copper electrodes up to 8 K. For the measurements on powders, the SWCNTs were filled into a double T-shaped sample consisting of a PVDF tube (inner diameter 3.8 mm, length 16 mm) sealed with copper plugs (see Figure 1 in [57]). The measurement of voltage and resistance was performed using the Keithley multimeter DMM2001 (Keithley Instruments, Cleveland, OH, USA) as a 4-wire technique for powders. The values given represent the mean values of three measurements.

3. Results

3.1. Dispersion Stability

The stability of the aqueous SWCNT dispersions was observed over a longer period of time (Figure 2). It can be clearly seen that the oxidation of the SWCNTs leads to a significantly more homogeneous and stable SWCNT dispersion. For the as-grown SWCNTs, agglomerates can be seen at the bottom of the glass while the solution remains clear. The fine dispersion of SWCNT-ox appears dark grey and remains like this for 24 h. The dispersions containing PDA already appear grey-brown due to the PDA (samples 3 and 4), but the dispersion with SWCNT-ox is darker than the dispersion with as-grown SWCNT and also stable over a longer time. It is known from the literature that the oxidation of CNTs leads to the formation of polar groups on the surface, followed by a significant increase in their polarity, which thus improves their dispersibility in polar water [63–65]. From the stability test, it can be concluded that the PDA coating of oxidized SWCNTs leads to the best SWCNT distribution in polar medium water, which is advantageous for the next process step of nickel deposition.

The high dispersion stability of both SWCNT types coated with PDA is a prerequisite for good accessibility of the SWCNT surface in the metallization process. While functionalization with an organic material does not change their properties much, metallization creates a dense nickel shell (Ni density is about 8.9 g/cm^3). In comparison, the density of plain SWCNTs is assumed to be around 1.5 g/cm^3 for SWCNTs with an outer diameter of 2 nm [66]. Consequently, the properties of the nanomaterials, i.e., surface/volume ratio and density, as well as dispersion and sedimentation behaviour are significantly affected by the nickel shell (compare Figure 1). Nickel-coated SWCNTs settle to the bottom as sediment.

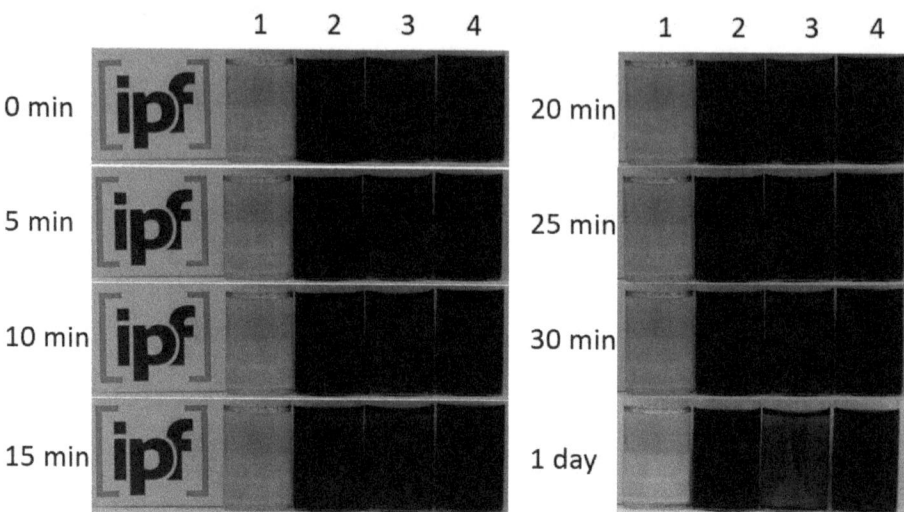

Figure 2. Sedimentation stability over one day. Photographs of aqueous dispersion of SWCNT (1), SWCNT-ox (2), SWCNT/PDA (3), SWCNT-ox/PDA (4).

3.2. Morphological Characterisation

In Figure 3, the SEM images of the differently modified SWCNTs are shown. In the case of the as-grown SWCNTs, the PDA coating (Figure 3b) hardly leads to any change in the appearance of the SWCNTs (Figure 3a). In the SWCNT-ox/PDA sample (Figure 3d), the SWCNTs appear thicker and more separated than in the SWCNT/PDA sample (Figure 3b). After Ni coating, the SWCNT/PDA/Ni appears like a glued mat of SWCNTs (Figure 3c). In contrast, the SWCNT-ox/PDA/Ni (Figure 3e) can be recognized as individual fibres that are significantly thicker than the SWCNT-ox/PDA. Also, some spherical particles are visible. In summary, the coating of the oxidized SWCNT with PDA and nickel resulted in a more homogeneous material.

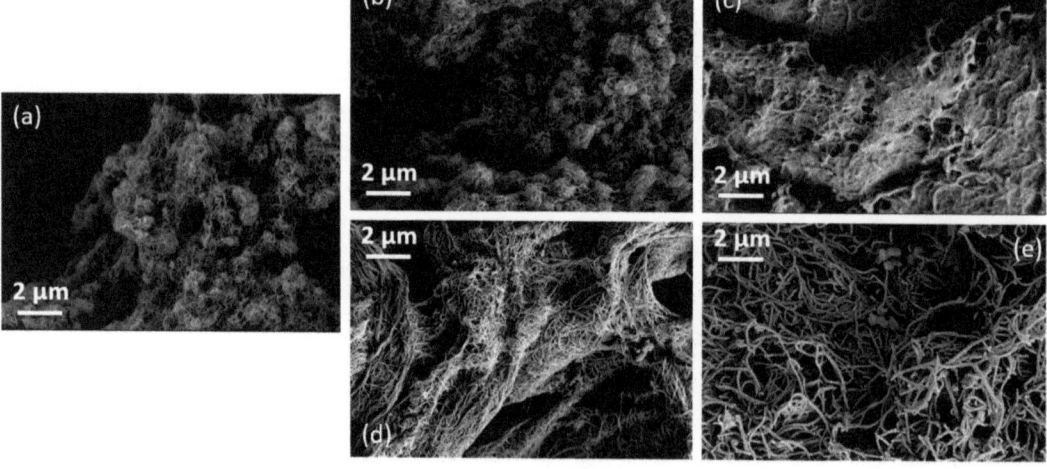

Figure 3. SEM image of SWCNT unmodified (**a**) SWCNT/PDA (**b**), SWCNT/PDA/Ni (**c**), and SWCNT-ox /PDA (**d**), SWCNT-ox/PDA/Ni (**e**).

In Figure 4a,b, the TEM images of both SWCNT types coated with PDA are shown. It is assumed that the black spherical particles in the images are PDA agglomerates from the aqueous dispersion. It seems that the PDA particles preferentially attach to the oxidized SWCNTs, while in the case of the non-functionalized SWCNTs, the PDA particles tend to lie next to the SWCNTs. The thin PDA coating cannot be reliably detected for any SWCNT type, even at higher magnifications. The TEM images of both SWCNT types coated with PDA and Ni (Figure 4c,d) show small grey areas adhering to the SWCNTs. It is assumed that this is nickel. For both SWCNT types, nickel always seems to be associated with the SWCNTs. Whether the SWCNTs were evenly coated cannot be estimated from the images.

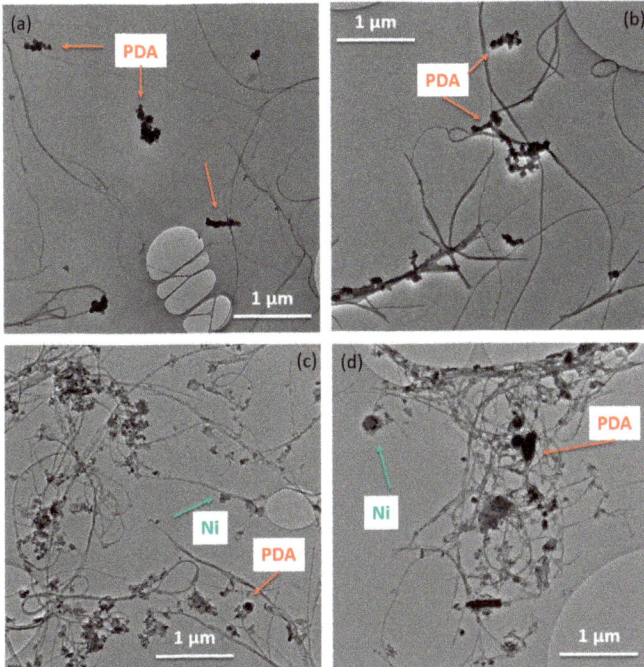

Figure 4. TEM image of SWCNT/PDA (**a**) and SWCNT-ox /PDA (**b**), SWCNT/PDA/Ni (**c**), SWCNT-ox/PDA/Ni (**d**).

Figure 5 shows AFM images of individual pristine, oxidized, and PDA-coated SWCNT. A comparison of pristine and oxidized CNTs is difficult because the effect of the oxidation is on the atomic scale and therefore below the resolution of the AFM. The PDA layer is hardly visible on the non-oxidized SWCNTs, neither in the topography (Figure 5b) nor in the phase image (Figure 5c). On a contrary, topography and phase images of the PDA-coated oxidized SWCNTs (Figure 5e,f) clearly show the characteristic globular structure of the PDA layer.

In Figure 6, AFM images of PDA- and Ni-coated SWCNTs are compared. Besides the height images, amplitude error images are presented. They show the first deviation of the height and, therefore, small structures. It is clearly visible that on oxidized fibres with a more homogeneous PDA coating (Figure 6b,d) the Ni distribution is more uniform, while on unoxidized PDA-coated SWCNTs (Figure 6a,c) the Ni is arranged in clusters.

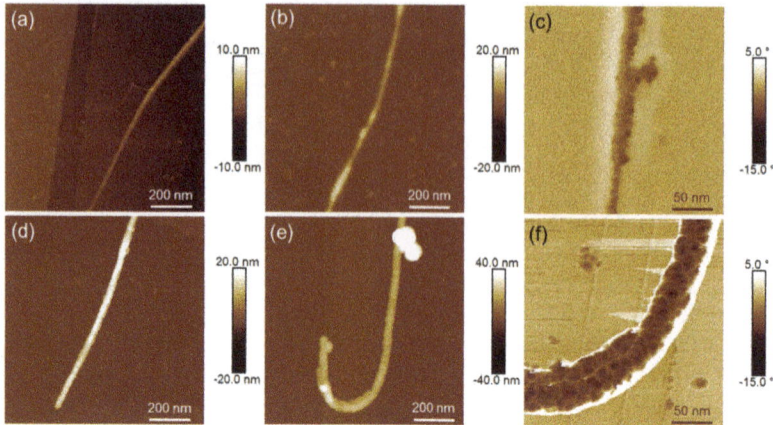

Figure 5. Topography images of individual SWCNTs: SWCNT (**a**), SWCNT-ox (**d**), SWCNT/PDA (**b**), SWCNT-ox/PDA (**e**); phase images of PDA-modified SWCNTs at higher magnification: SWCNT/PDA (**c**), SWCNT-ox/PDA (**f**).

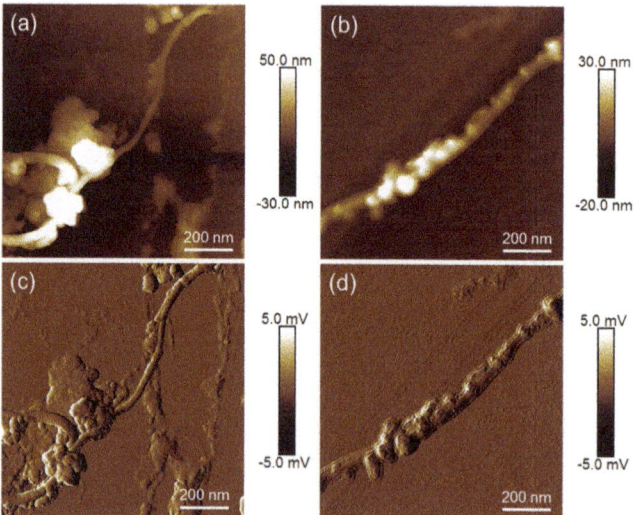

Figure 6. Topography images (**a,b**) and amplitude error images (**c,d**) of metallized SWCNTs: SWCNT/PDA/Ni (**a,d**), SWCNT-ox/PDA/Ni (**b,d**).

3.3. Investigation of the Chemical Composition and Bonding States near the Surface

As can be seen in the XPS wide-scan spectrum, the pristine SWCNT sample contains only traces of oxygen ([O]:[C] = 0.019) (Figure 7a). The majority of this oxygen is probably bound to iron, which was also clearly detected in the sample as Fe 2s, Fe 2p, and Fe LMM Auger series. No component peaks for oxidized carbon species could be separated in the high-resolution C 1s spectrum (Figure 8a). The main component peak *Gr* in the C 1s spectrum at 283.99 eV results from photoelectrons of the sp^2-hybridized carbon atoms of the graphite-like lattices in their electronic ground states. Photoelectrons from electronically exited states produced by $\pi \rightarrow \pi^*$ transitions were collected as shake-up peaks on the spectrum's high-energy side.

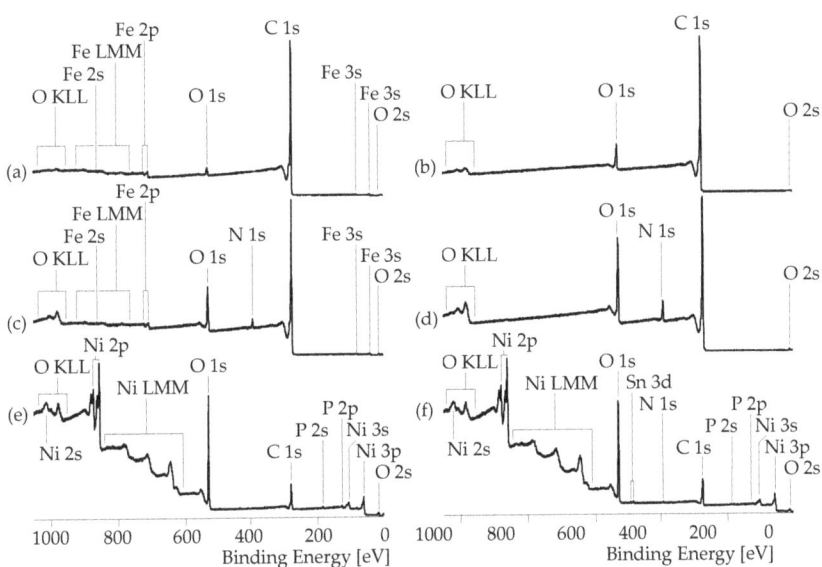

Figure 7. XPS wide-scan spectra recorded from pristine SWCNT (**a**), SWCNT-ox (**b**), SWCNT/PDA (**c**), SWCNT-ox/PDA (**d**), SWCNT/PDA/Ni (**e**), and SWCNT-ox/PDA/Ni (**f**) samples.

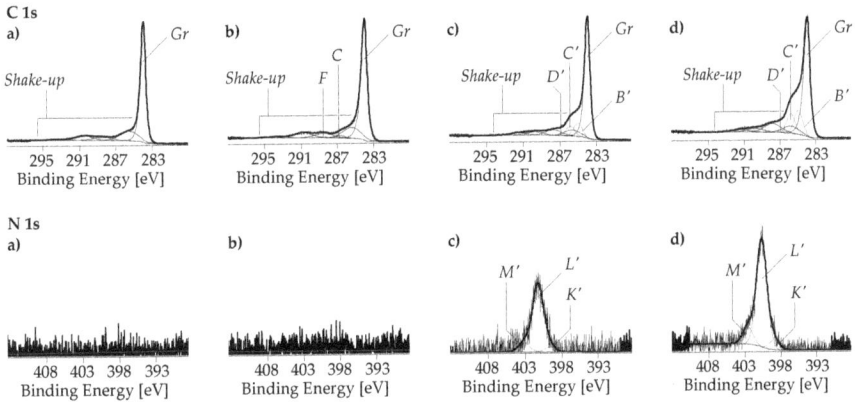

Figure 8. C 1s and N 1s high-resolution XPS spectra recorded from pristine SWCNT (**a**), SWCNT-ox (**b**), SWCNT/PDA (**c**), and SWCNT-ox/PDA samples. The blue component peaks in the figures (**c**) and (**d**) demonstrate the presence of PDA on the SWCNT surfaces (their origins).

The oxidation of the SWCNTs increased not only the relative oxygen content significantly to [O]:[C] = 0.056, but it also removed the iron impurities in the sample (Figure 7b). The component peaks C (at 286.54 eV) and F (at 288.33 eV) separated in the C 1s spectrum (Figure 8b) showed that oxidation of the carbon took place and that mainly phenolic C–OH groups (component peak C) and carboxylic acid groups (component peak F) were introduced in the surfaces of the SWCNT-ox sample.

Deposition of PDA creates N 1s peaks and enhances the oxygen peaks in the spectra. Taking its intensity as a measure of the PDA content in the surface region of the SWCNTs, it can be seen that the oxidation had a beneficial effect on the adsorption and interfacial polymerization of the DA (Figure 7c,d). It increased the relative nitrogen content from [N]:[C] = 0.036 for the SWCNT/PDA sample to [N]:[C] = 0.06 for the SWCNT-ox/PDA sample, i.e., about twice the amount of PDA was deposited on the hydrophilic SWCNT-ox surface. The corresponding C 1s and N 1s spectra (Figure 8c,d) show the characteristic component peaks indicating the presence of PDA on the SWCNT surfaces. In addition to the main component peaks Gr and the shake-up peaks, we found the component peaks B' (at ca. 285.11 eV), C' (at ca. 285.97 eV), and D' (at ca. 287.25 eV). Component peaks B', which have twice the intensity as the [N]:[C] ratios, show the C–N bonds of the PDA's indoline and indole units. Component peaks C' result from the catechol groups ($^{C'}\underline{C}$–OH). Since some of the catechol groups were present in their oxidized form, namely as quinone-like groups, their intensities were slightly smaller than the intensities of peaks B'. The quinone-like groups ($^{D'}\underline{C}$=O) were represented by the component peaks D'. The main component peaks L' (at ca. 399.7 eV) in the N 1s spectra show the fractions of the PDA's indoline-like bonded nitrogen atoms (C–$^{L'}\underline{N}$H–C). Due to the increased electron density of the indole-bonded nitrogen atoms (C=$^{K'}\underline{N}$–C ↔ C–$^{K'}\underline{N}$=C), the component peaks K' were shifted to lower binding energy values (ca. 398.03 eV). Protonated secondary amino groups (C–$^{M'}\underline{N}^{\oplus}H_2$–C) were identified as component peaks M' at ca. 401.58 eV.

After the metallization, the element peaks of nickel (Ni 2s, Ni 2p, and Ni LMM *Auger* series) appear in the wide-scan spectra (Figure 7e,f). Surprisingly, the relative nickel content of SWCNT/PDA with lower PDA content was with [Ni]:[C] = 0.509, significantly greater than that of the SWCNT-ox/PDA sample with more PDA in the SWCNT/nickel interphase ([Ni]:[C] = 0.305). This is surprising at first glance but can be explained by the different distribution of the nickel visible in the SEM images (Figure 3). On the SWCNT/PDA/Ni sample, a lot of metal is deposited between the CNTs, while on the SWCNT-ox/PDA/Ni sample, only the CNTs are covered with nickel.

The almost complete disappearance of the N 1s peaks together with the changed shapes of the C 1s spectra (Figure 9a,b) shows that the PDA film is mostly covered by the nickel layer. The presence of carbon is now largely due to the presence of surface contaminants, such as fatty acid esters that spontaneously adsorb onto metal oxides and minimize their surface-free energy. Detailed analysis shows, however, that the nickel layers are not fully closed. In the deconvolution process of the spectra, it was necessary to introduce the component peak Gr (at ca. 284.3 eV), which indicates the presence of graphite-like bonded carbon atoms in the surface region. The intense component peaks A'' (at 285.00 eV) result from photoelectrons from sp^3-hybridized carbon atoms that have no heteroatoms in their immediate vicinity. Carboxylic ester groups are identified by the component peaks C'' (alcohol-sided carbon atoms, O=C–O–$^C\underline{C}$) and E'' (carbonyl carbon atoms, O=$^E\underline{C}$–O–C). Component peaks F'' arise from carboxylic acid groups (O=$^{F''}\underline{C}$–OH) and their corresponding carboxylates (O=$^{F''}\underline{C}$–O$^\ominus$ ↔ $^\ominus$O–$^{F''}\underline{C}$=O). Carbon atoms in α-position to the carbonyl carbon atoms ($^{B''}\underline{C}$–COO) are observed as component peaks B''. The small component peaks D'' at 287.7 eV are an indication of the presence of ketone groups (O=$^{D''}\underline{C}$).

The complex shapes of the Ni 2p spectra (Figure 9c,d) and the high relative oxygen contents (Figure 7e,f) indicate that nickel is present preferentially in its oxidized forms (formally as Ni^{2+}). The positions of the main component peaks U'' in the high-resolution Ni 2p$_{3/2}$ spectra at ca. 855.64 eV confirm the presence of $\underline{Ni}$O and/or $\underline{Ni}$(OH)$_2$ [67]. The small component peaks T'' on the low-energy side of the main component peak result from metallic nickel ($\underline{Ni}^0$). The divalent nature of the nickel ions (Ni^{2+}) mainly present in the samples corresponds to the [Ar]4s^{0}3d^8 electron configuration with a magnetic moment of 2.83 Bohr magneton. The photoionization of this electron configuration results in the [1s^{2}2s^{2}2p^6–^{1}3s^{2}3p^6]4s^{0}3d^8 electron state, which can be neutralized by electrons from the oxygen-containing ligands (L) situated in the nickel's coordination sphere. The so-called *local screening* resulted in the [1s^{2}2s^{2}2p^6–^{1}3s^{2}3p^6]4s^{0}3d^9[L^{-1}] electron ground state causing the component peaks U'' mentioned above. An electronically excited state resulted from a second electron transfer from the ligands to Ni^{2+}. Photoelectrons from this [1s^{2}2s^{2}2p^6–^{1}3s^{2}3p^6] 4s^{0}3d^{10}[L^{-2}] state are observed as satellite peaks ($S2$) at ca. 861.35 eV. Oxygen-containing ligands that are further away can also provide electrons for the photoionized nickel ions via a transfer by hopping (non-local screening to the $[1s^22s^22p^{6-1}3s^23p^6]4s^03d^9[L][3d^8L^{-1}]$ state). The corresponding satellite peaks (*S1*) are found at ca. 857.46 eV. Finally, satellite peaks *S3* arising at ca. 863.57 result from photoelectrons in the $[Ar]4s^03d^8$ state, which are not neutralized by the electrons provided by the ligands.

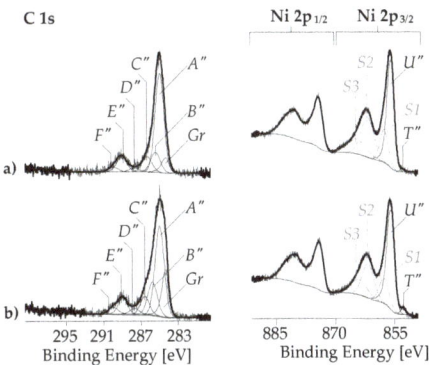

Figure 9. C 1s and Ni 2p high-resolution XPS spectra recorded from SWCNT/PDA/Ni (**a**), and SWCNT-ox/PDA/Ni (**b**) samples. (The origins of the component peaks are explained in the text).

3.4. Thermoelectric Characterisation

In Figure 10 the values of volume conductivity, Seebeck coefficient, and power factor are summarized for the three different modification states of SWCNT and oxidized SWCNT. The Seebeck coefficient of pristine SWCNT powder is determined to be 39.6 µV/K [57]. The oxidation of the SWCNT leads to a lower S-value of 22.4 ± 0.1 µV/K. It is assumed that the insertion of hydroxy and carboxyl groups partially destroys the perfect structure of the SWCNT walls and thus impairs electron transport across the SWCNT walls, resulting in a lower S-value. At the same time, SWCNT oxidation leads to an increase in volume conductivity from 1790 S/m [57] up to 3291 ± 59 S/m. Thus, the PF decreases from 2.8 [57] to 1.1 µW/m·K² due to the lower Seebeck coefficient.

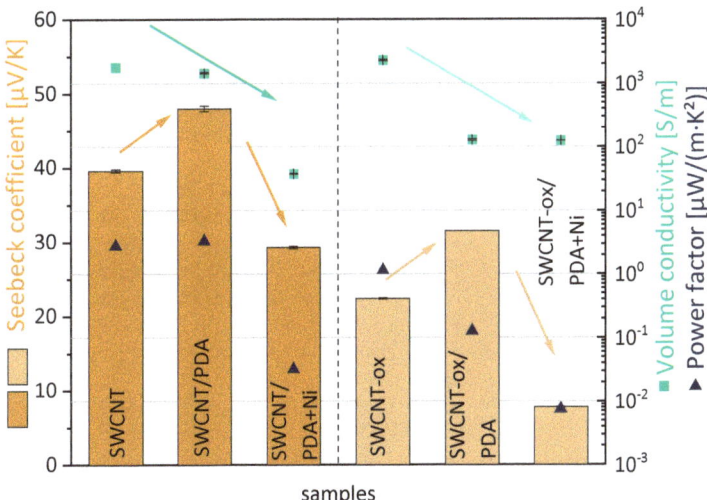

Figure 10. Thermoelectric parameter of SWCNTs at different modification state (the green and brown arrows only serve to illustrate the trends).

For both as-grown SWCNTs and oxidized SWCNTs, the PDA coating leads to an increase in the S-value by 9 µV/K, such that S-values of 48.0 ± 0.4 µV/K (SWCNT) and 31.5 ± 0.0 µV/K (SWCNT-ox) are achieved, i.e., PDA acts as a p-dopant agent. It is surprising that the effect is about the same for SWCNTs and SWCNT-ox because both morphological examination and XPS showed a thicker and more homogeneous PDA coating on the SWCNT-ox. In addition, after PDA coating, the electrical conductivity of SWCNT-ox powders drops sharply to 128 ± 4 S/m, and only slightly for SWCNTs (1451 ± 32 S/m). The electrically insulating PDA layer can presumably deteriorate electron transport by direct contact with the SWCNTs or even electron hopping. Thus, the PDA modification leads to a slightly higher PF for SWCNT/PDA at 3.3 µW/m·K^2 compared to SWCNT. However, for the SWCNT-ox, the PF decreases from 1.1 µW/m·K^2 (SWCNT-ox) to 0.1 µW/m·K^2 (SWCNT-ox/PDA) due to the strongly decreased volume conductivity.

The coating of both SWCNT/PDA types with nickel leads to a reduction of the Seebeck coefficient to 29.3 ± 0.1 µV/K (SWCNT) and 7.8 ± 0.0 µV/K (SWCNT-ox). This reduction is consistent with the negative S-value of the nickel itself of −19 µV/K [48,62]. Nickel thus acts as an n-dopant, as expected. This effect is clearly more pronounced with the oxidized SWCNTs. Here, the S-value is reduced by 23.8 µV/K instead of only 18.7 µV/K for SWCNT/PDA. Obviously the more uniform nickel coating of the oxidized SWCNTs, as seen in Figure 3, has a stronger effect on the S-value despite the lower amount of deposited nickel revealed by XPS. Unfortunately, the amount of nickel deposited was not sufficient to produce SWCNTs with a negative Seebeck coefficient. The XPS study showed that the nickel layer is not closed on the SWCNT surface. Thus, p- and n-conducting structures alternate. Presumably, this is one reason why no negative S-value was obtained. Thus, the PF decreases for both samples to 0.03 µW/m·K^2 (SWCNT) and 0.01 µW/m·K^2 (SWCNT-ox).

4. Discussion

The present study investigates how SWCNTs can be coated with nickel using PDA as an adhesive coupler in a "green" process in aqueous solutions. With regard to thermoelectric applications, the aim was to reverse the Seebeck coefficient of the SWCNTs to negative values since nickel as a metal has a negative S-value of −19 µV/K.

Oxidation of the SWCNTs improved their dispersibility in water. The better dispersibility of the oxidized SWCNTs remained after modification of the SWCNTs with PDA, which led to better modification results.

Both pristine and oxidized SWCNTs were modified with dopamine and metallized with nickel. XPS verified the deposition of PDA and nickel on both pristine and oxidized SWCNTs. While more PDA was detected on oxidized SWCNTs, the elemental ratio of nickel was significantly higher on non-oxidized SWCNTs. This corresponds to morphology studies of the SWCNTs carried out by SEM, TEM, and AFM. On the surfaces of oxidized SWCNTs, PDA and nickel are homogeneously distributed. On non-oxidized SWCNTs, however, no continuous PDA layer was detected, and nickel was deposited not only on the CNTs but also in the spaces between them.

The thermoelectric characterisation shows that PDA acts as a p-dopant, resulting in an increase in the Seebeck coefficient by around 9 µV/K for both pristine and oxidized SWCNT. This increase was not helpful with regard to the aim of the study (reduction of the S-value), but the PDA coating was only an intermediate step. The coating with nickel successfully reduced the S-value by 19 µV/K (SWCNT/PDA) and 23.8 µV/K (oxidized SWCNT/PDA). The deposited amount was, however, not sufficient to create a negative Seebeck coefficient.

The results show that electroless metallization is able to create a nickel layer on PDA-modified SWCNTs. Prior oxidation of the SWCNTs is essential to ensuring good dispersion and homogeneous coating with PDA and nickel. The nickel coating can actually change the conduction type of the SWCNTs, but more work is needed to adjust the nickel layer thickness.

In future studies, the thermoelectric properties will be investigated as a function of Ni content or Ni layer thickness on the SWCNT. On the other hand, it is interesting that the PDA coating increases the S-value. Here, too, the dependence between the PDA coverage of the SWCNTs and their thermoelectric properties is worth studying.

Author Contributions: Conceptualization, C.Z. and B.K.; methodology, C.Z. and B.K.; validation, F.S., B.K., C.Z., A.D. and A.J.; investigation, S.P., F.S., A.J. and A.D.; resources, C.Z. and B.K.; writing—original draft preparation, C.Z. and B.K.; writing—review and editing, F.S. and A.D.; visualization, C.Z. and B.K.; project administration, C.Z. and B.K. All authors have read and agreed to the published version of the manuscript.

Funding: This research received no external funding.

Data Availability Statement: The data presented in this study are available upon request from the corresponding author.

Acknowledgments: The authors would like to thank the collaborators of the IPF research technology department for their support, Marie Mokbel for dispersion experiments and metallization, Ulrike Jentzsch-Hutschenreuther for the thermoelectric measurements and Manuela Heber (both IPF) for sample preparation for SEM and TEM imaging.

Conflicts of Interest: The authors declare no conflict of interest.

References

1. Fiedler, H.; Toader, M.; Hermann, S.; Rennau, M.; Rodriguez, R.D.; Sheremet, E.; Hietschold, M.; Zahn, D.R.T.; Schulz, S.E.; Gessner, T. Back-end-of-line compatible contact materials for carbon nanotube based interconnects. *Microelectron. Eng.* **2015**, *137*, 130–134. [CrossRef]
2. Dorri Moghadam, A.; Omrani, E.; Menezes, P.L.; Rohatgi, P.K. Mechanical and tribological properties of self-lubricating metal matrix nanocomposites reinforced by carbon nanotubes (CNTs) and graphene—A review. *Compos. Part B* **2015**, *77*, 402–420. [CrossRef]
3. Alim, M.A.; Abdullah, M.Z.; Aziz, M.S.A.; Kamarudin, R.; Gunnasegaran, P. Recent Advances on Thermally Conductive Adhesive in Electronic Packaging: A Review. *Polymers* **2021**, *13*, 3337. [CrossRef]
4. Boddu, V.M.; Brenner, M.W. Energy dissipation in intercalated carbon nanotube forests with metal layers. *Appl. Phys. A* **2016**, *122*, 88. [CrossRef]
5. Selvakumar, A.; Perumalraj, R.; Jeevananthan, P.N.R.; Archana, S.; Sudagar, J. Electroless NiP–MWCNT composite coating for textile industry application. *Surf. Eng.* **2016**, *32*, 338–343. [CrossRef]
6. Arai, S.; Osaki, T.; Hirota, M.; Uejima, M. Fabrication of copper/single-walled carbon nanotube composite film with homogeneously dispersed nanotubes by electroless deposition. *Mater. Today Commun.* **2016**, *7*, 101–107. [CrossRef]
7. Lai, X.; Guo, R.; Xiao, H.; Lan, J.; Jiang, S.; Cui, C.; Qin, W. Flexible conductive copper/reduced graphene oxide coated PBO fibers modified with poly(dopamine). *J. Alloys Compd.* **2019**, *788*, 1169–1176. [CrossRef]
8. Alipour Ghorbani, N.; Namazi, H. Polydopamine-graphene/Ag–Pd nanocomposite as sustainable catalyst for reduction of nitrophenol compounds and dyes in environment. *Mater. Chem. Phys.* **2019**, *234*, 38–47. [CrossRef]
9. Jiang, Y.; Lu, Y.; Zhang, L.; Liu, L.; Dai, Y.; Wang, W. Preparation and characterization of silver nanoparticles immobilized on multi-walled carbon nanotubes by poly(dopamine) functionalization. *J. Nanopart. Res.* **2012**, *14*, 938. [CrossRef]
10. Zhu, C.; Guan, X.; Wang, X.; Li, Y.; Chalmers, E.; Liu, X. Mussel-Inspired Flexible, Durable, and Conductive Fibers Manufacturing for Finger-Monitoring Sensors. *Adv. Mater. Interfaces* **2019**, *6*, 1801547. [CrossRef]
11. Djokić, S.S. Electroless Deposition of Metals and Alloys. In *Modern Aspects of Electrochemistry*; Conway, B.E., White, R.E., Eds.; Springer: Boston, MA, USA, 2002; pp. 51–133. [CrossRef]
12. Hu, Q.-h.; Wang, X.-t.; Chen, H.; Wang, Z.-f. Synthesis of Ni/graphene sheets by an electroless Ni-plating method. *New Carbon Mater.* **2012**, *27*, 35–41. [CrossRef]
13. Rowe, D.M. *CRC Handbook of Thermoelectrics*; CRC Press: Boca Raton, FL, USA, 1995.
14. Piao, M.; Alam, M.R.; Kim, G.; Dettlaff-Weglikowska, U.; Roth, S. Effect of chemical treatment on the thermoelectric properties of single walled carbon nanotube networks. *Phys. Status Solidi B* **2012**, *249*, 2353–2356. [CrossRef]
15. Nonoguchi, Y.; Ohashi, K.; Kanazawa, R.; Ashiba, K.; Hata, K.; Nakagawa, T.; Adachi, C.; Tanase, T.; Kawai, T. Systematic Conversion of Single Walled Carbon Nanotubes into n-type Thermoelectric Materials by Molecular Dopants. *Sci. Rep.* **2013**, *3*, 3344. [CrossRef] [PubMed]
16. Tzounis, L.; Hegde, M.; Liebscher, M.; Dingemans, T.; Pötschke, P.; Paipetis, A.S.; Zafeiropoulos, N.E.; Stamm, M. All-aromatic SWCNT-Polyetherimide nanocomposites for thermal energy harvesting applications. *Compos. Sci. Technol.* **2018**, *156*, 158–165. [CrossRef]

17. Mytafides, C.K.; Tzounis, L.; Karalis, G.; Formanek, P.; Paipetis, A.S. High-Power All-Carbon Fully Printed and Wearable SWCNT-Based Organic Thermoelectric Generator. *ACS Appl. Mater. Interfaces* **2021**, *13*, 11151–11165. [CrossRef]
18. Hata, S.; Maeshiro, K.; Shiraishi, M.; Du, Y.; Shiraishi, Y.; Toshima, N. Surfactant-Wrapped n-Type Organic Thermoelectric Carbon Nanotubes for Long-Term Air Stability and Power Characteristics. *ACS Appl. Electron. Mater.* **2022**, *4*, 1153–1162. [CrossRef]
19. Krause, B.; Imhoff, S.; Voit, B.; Pötschke, P. Influence of Polyvinylpyrrolidone on Thermoelectric Properties of Melt-Mixed Polymer/Carbon Nanotube Composites. *Micromachines* **2023**, *14*, 181. [CrossRef]
20. Hongmei, W.; Jun, Y.; Zhou, D. Review of Recent Developments in Thermoelectric Materials. In Proceedings of the 2016 International Conference on Robots & Intelligent System (ICRIS), ZhangJiaJie, China, 27–28 August 2016; pp. 394–397.
21. Nolas, G.S.; Sharp, J.; Goldsmid, H. *Thermoelectrics: Basic Principles and New Materials Developments*; Springer: Berlin/Heidelberg, Germany, 2001.
22. Goldsmid, H.J. *Introduction to Thermoelectricity*; Springer: Berlin/Heidelberg, Germany, 2010. [CrossRef]
23. Kurkowska, M.; Awietjan, S.; Kozera, R.; Jezierska, E.; Boczkowska, A. Application of electroless deposition for surface modification of the multiwall carbon nanotubes. *Chem. Phys. Lett.* **2018**, *702*, 38–43. [CrossRef]
24. Melzer, M.; Waechtler, T.; Müller, S.; Fiedler, H.; Hermann, S.; Rodriguez, R.D.; Villabona, A.; Sendzik, A.; Mothes, R.; Schulz, S.E.; et al. Copper oxide atomic layer deposition on thermally pretreated multi-walled carbon nanotubes for interconnect applications. *Microelectron. Eng.* **2013**, *107*, 223–228. [CrossRef]
25. Maheswaran, R.; Shanmugavel, B.P. A Critical Review of the Role of Carbon Nanotubes in the Progress of Next-Generation Electronic Applications. *J. Electron. Mater.* **2022**, *51*, 2786–2800. [CrossRef]
26. Lee, H.; Dellatore, S.M.; Miller, W.M.; Messersmith, P.B. Mussel-Inspired Surface Chemistry for Multifunctional Coatings. *Science* **2007**, *318*, 426–430. [CrossRef]
27. Barclay, T.G.; Hegab, H.M.; Clarke, S.R.; Ginic-Markovic, M. Versatile Surface Modification Using Polydopamine and Related Polycatecholamines: Chemistry, Structure, and Applications. *Adv. Mater. Interfaces* **2017**, *4*, 1601192. [CrossRef]
28. Augustine, N.; Putzke, S.; Janke, A.; Simon, F.; Drechsler, A.; Zimmerer, C.A. Dopamine-Supported Metallization of Polyolefins—A Contribution to Transfer to an Eco-friendly and Efficient Technological Process. *ACS Appl. Mater. Interfaces* **2022**, *14*, 5921–5931. [CrossRef] [PubMed]
29. Liebscher, J.; Mrówczyński, R.; Scheidt, H.A.; Filip, C.; Hădade, N.D.; Turcu, R.; Bende, A.; Beck, S. Structure of Polydopamine: A Never-Ending Story? *Langmuir* **2013**, *29*, 10539–10548. [CrossRef]
30. Ho, C.C.; Ding, S.J. Structure, properties and applications of mussel-inspired polydopamine. *J. Biomed. Nanotechnol.* **2014**, *10*, 3063–3084. [CrossRef] [PubMed]
31. Benko, A.; Duch, J.; Gajewska, M.; Marzec, M.; Bernasik, A.; Nocuń, M.; Piskorz, W.; Kotarba, A. Covalently bonded surface functional groups on carbon nanotubes: From molecular modeling to practical applications. *Nanoscale* **2021**, *13*, 10152–10166. [CrossRef] [PubMed]
32. Tsuji, Y.; Yoshizawa, K. Competition between Hydrogen Bonding and Dispersion Force in Water Adsorption and Epoxy Adhesion to Boron Nitride: From the Flat to the Curved. *Langmuir* **2021**, *37*, 11351–11364. [CrossRef]
33. Nakamura, S.; Tsuji, Y.; Yoshizawa, K. Role of Hydrogen-Bonding and OH−π Interactions in the Adhesion of Epoxy Resin on Hydrophilic Surfaces. *ACS Omega* **2020**, *5*, 26211–26219. [CrossRef]
34. Stafiej, A.; Pyrzynska, K. Adsorption of heavy metal ions with carbon nanotubes. *Sep. Purif. Technol.* **2007**, *58*, 49–52. [CrossRef]
35. Hsieh, S.-H.; Horng, J.-J. Adsorption behavior of heavy metal ions by carbon nanotubes grown on microsized Al_2O_3 particles. *J. Univ. Sci. Technol. Beijing Miner. Metall. Mater.* **2007**, *14*, 77–84. [CrossRef]
36. Poorsargol, M.; Razmara, Z.; Amiri, M.M. The role of hydroxyl and carboxyl functional groups in adsorption of copper by carbon nanotube and hybrid graphene–carbon nanotube: Insights from molecular dynamic simulation. *Adsorption* **2020**, *26*, 397–405. [CrossRef]
37. Fei, B.; Qian, B.; Yang, Z.; Wang, R.; Liu, W.C.; Mak, C.L.; Xin, J.H. Coating carbon nanotubes by spontaneous oxidative polymerization of dopamine. *Carbon* **2008**, *46*, 1795–1797. [CrossRef]
38. Wang, Q.; Callisti, M.; Miranda, A.; McKay, B.; Deligkiozi, I.; Milickovic, T.K.; Zoikis-Karathanasis, A.; Hrissagis, K.; Magagnin, L.; Polcar, T. Evolution of structural, mechanical and tribological properties of Ni–P/MWCNT coatings as a function of annealing temperature. *Surf. Coat. Technol.* **2016**, *302*, 195–201. [CrossRef]
39. Wang, Y.; Chen, J. Preparation and Characterization of Polydopamine-Modified Ni/Carbon Nanotubes Friction Composite Coating. *Coatings* **2019**, *9*, 596. [CrossRef]
40. Mondin, G.; Wisser, F.M.; Leifert, A.; Mohamed-Noriega, N.; Grothe, J.; Dörfler, S.; Kaskel, S. Metal deposition by electroless plating on polydopamine functionalized micro- and nanoparticles. *J. Colloid Interface Sci.* **2013**, *411*, 187–193. [CrossRef] [PubMed]
41. Choi, J.-R.; Lee, Y.S.; Park, S.-J. A study on thermal conductivity of electroless Ni–B plated multi-walled carbon nanotubes-reinforced composites. *J. Ind. Eng. Chem.* **2014**, *20*, 3421–3424. [CrossRef]
42. Park, C.; Kim, T.; Samuel, E.P.; Kim, Y.-I.; An, S.; Yoon, S.S. Superhydrophobic antibacterial wearable metallized fabric as supercapacitor, multifunctional sensors, and heater. *J. Power Sources* **2021**, *506*, 230142. [CrossRef]
43. Ang, L.M.; Hor, T.S.A.; Xu, G.Q.; Tung, C.H.; Zhao, S.P.; Wang, J.L.S. Decoration of activated carbon nanotubes with copper and nickel. *Carbon* **2000**, *38*, 363–372. [CrossRef]

44. Zhai, J.; Cui, C.; Ren, E.; Zhou, M.; Guo, R.; Xiao, H.; Li, A.; Jiang, S.; Qin, W. Facile synthesis of nickel/reduced graphene oxide-coated glass fabric for highly efficient electromagnetic interference shielding. *J. Mater. Sci. Mater. Electron.* **2020**, *31*, 8910–8922. [CrossRef]
45. Popescu, S.M.; Barlow, A.J.; Ramadan, S.; Ganti, S.; Ghosh, B.; Hedley, J. Electroless Nickel Deposition: An Alternative for Graphene Contacting. *ACS Appl. Mater. Interfaces* **2016**, *8*, 31359–31367. [CrossRef]
46. Zhai, T.; Di, L.; Yang, D. Study on the Pretreatment of Poly(ether ether ketone)/Multiwalled Carbon Nanotubes Composites through Environmentally Friendly Chemical Etching and Electrical Properties of the Chemically Metallized Composites. *ACS Appl. Mater. Interfaces* **2013**, *5*, 12499–12509. [CrossRef]
47. Arai, S.; Fujimori, A.; Murai, M.; Endo, M. Excellent solid lubrication of electrodeposited nickel-multiwalled carbon nanotube composite films. *Mater. Lett.* **2008**, *62*, 3545–3548. [CrossRef]
48. Burkov, A.T.; Heinrich, A.; Konstantinov, P.P.; Nakama, T.; Yagasaki, K. Experimental set-up for thermopower and resistivity measurements at 100–1300 K. *Meas. Sci. Technol.* **2001**, *12*, 264–272. [CrossRef]
49. Xu, C.; Wu, G.; Liu, Z.; Wu, D.; Meek, T.T.; Han, Q. Preparation of copper nanoparticles on carbon nanotubes by electroless plating method. *Mater. Res. Bull.* **2004**, *39*, 1499–1505. [CrossRef]
50. Hanna, F.; Hamid, Z.A.; Aal, A.A. Controlling factors affecting the stability and rate of electroless copper plating. *Mater. Lett.* **2004**, *58*, 104–109. [CrossRef]
51. Gui, C.; Yao, C.; Huang, J.; Chen, Z.; Yang, G. Preparation of polymer brush/Ni particle and its application in electroless copper plating on PA12 powder. *Appl. Surf. Sci.* **2020**, *506*, 144935. [CrossRef]
52. Dejene, F.K.; Flipse, J.; van Wees, B.J. Spin-dependent Seebeck coefficients of Ni80Fe20 and Co in nanopillar spin valves. *Phys. Rev. B* **2012**, *86*, 024436. [CrossRef]
53. Hao, M.; Tang, M.; Wang, W.; Tian, M.; Zhang, L.; Lu, Y. Silver-nanoparticle-decorated multiwalled carbon nanotubes prepared by poly(dopamine) functionalization and ultraviolet irradiation. *Compos. Part B* **2016**, *95*, 395–403. [CrossRef]
54. Vedernikov, M.V. The thermoelectric powers of transition metals at high temperature. *Adv. Phys.* **1969**, *18*, 337–370. [CrossRef]
55. Rowell, M.W.; Topinka, M.A.; McGehee, M.D.; Prall, H.-J.; Dennler, G.; Sariciftci, N.S.; Hu, L.; Gruner, G. Organic solar cells with carbon nanotube network electrodes. *Appl. Phys. Lett.* **2006**, *88*, 233506. [CrossRef]
56. Predtechenskiy, M.R.; Khasin, A.A.; Bezrodny, A.E.; Bobrenok, O.F.; Dubov, D.Y.; Muradyan, V.E.; Saik, V.O.; Smirnov, S.N. New perspectives in SWCNT applications: Tuball SWCNTs. Part 1. Tuball by itself—All you need to know about it. *Carbon Trends* **2022**, *8*, 100175. [CrossRef]
57. Krause, B.; Barbier, C.; Levente, J.; Klaus, M.; Pötschke, P. Screening of different carbon nanotubes in melt-mixed polymer composites with different polymer matrices for their thermoelectric properties. *J. Compos. Sci.* **2019**, *3*, 106. [CrossRef]
58. Krause, B.; Kroschwald, L.; Pötschke, P. The Influence of the Blend Ratio in PA6/PA66/MWCNT Blend Composites on the Electrical and Thermal Properties. *Polymers* **2019**, *11*, 122. [CrossRef] [PubMed]
59. Shirley, D.A. High-Resolution X-ray Photoemission Spectrum of the Valence Bands of Gold. *Phys. Rev. B* **1972**, *5*, 4709–4714. [CrossRef]
60. Paiva, M.C.; Simon, F.; Novais, R.M.; Ferreira, T.; Proença, M.F.; Xu, W.; Besenbacher, F. Controlled Functionalization of Carbon Nanotubes by a Solvent-free Multicomponent Approach. *ACS Nano* **2010**, *4*, 7379–7386. [CrossRef]
61. Magonov, S.N.; Elings, V.; Whangbo, M.H. Phase imaging and stiffness in tapping-mode atomic force microscopy. *Surf. Sci.* **1997**, *375*, L385–L391. [CrossRef]
62. Jenschke, W.; Ullrich, M.; Krause, B.; Pötschke, P. Messanlage zur Untersuchung des Seebeck-Effektes in Polymermaterialien—Measuring apparatus for study of Seebeck-effect in polymer materials. *Tech. Mess.* **2020**, *87*, 495–503. [CrossRef]
63. Lavagna, L.; Nisticò, R.; Musso, S.; Pavese, M. Functionalization as a way to enhance dispersion of carbon nanotubes in matrices: A review. *Mater. Today Chem.* **2021**, *20*, 100477. [CrossRef]
64. Xia, T.; Guo, X.; Lin, Y.; Xin, B.; Li, S.; Yan, N.; Zhu, L. Aggregation of oxidized multi-walled carbon nanotubes: Interplay of nanomaterial surface O-functional groups and solution chemistry factors. *Environ. Pollut.* **2019**, *251*, 921–929. [CrossRef]
65. Rajendran, D.; Ramalingame, R.; Adiraju, A.; Nouri, H.; Kanoun, O. Role of Solvent Polarity on Dispersion Quality and Stability of Functionalized Carbon Nanotubes. *J. Compos. Sci.* **2022**, *6*, 26. [CrossRef]
66. Laurent, C.; Flahaut, E.; Peigney, A. The weight and density of carbon nanotubes versus the number of walls and diameter. *Carbon* **2010**, *48*, 2994–2996. [CrossRef]
67. Naumkin, A.V.; Kraut-Vass, A.; Gaarenstroom, S.W.; Powell, C.J. NIST X-ray Photoelectron Spectroscopy Database. In *NIST Standard Reference Database 20, Version 4.1*; Measurement Services Division of the National Institute of Standards and Technology (NIST) Material Measurement Laboratory (MML): Gaithersburg, MD, USA, 2012. [CrossRef]

Disclaimer/Publisher's Note: The statements, opinions and data contained in all publications are solely those of the individual author(s) and contributor(s) and not of MDPI and/or the editor(s). MDPI and/or the editor(s) disclaim responsibility for any injury to people or property resulting from any ideas, methods, instructions or products referred to in the content.

 *nanomaterials*

Article

Realizing the Ultralow Lattice Thermal Conductivity of Cu$_3$SbSe$_4$ Compound via Sulfur Alloying Effect

Lijun Zhao [1], Haiwei Han [1], Zhengping Lu [1], Jian Yang [2,*], Xinmeng Wu [1], Bangzhi Ge [3], Lihua Yu [1], Zhongqi Shi [4], Abdulnasser M. Karami [5], Songtao Dong [1], Shahid Hussain [2], Guanjun Qiao [2] and Junhua Xu [1,*]

1. School of Materials Science and Engineering, Jiangsu University of Science and Technology, Zhenjiang 212100, China
2. School of Materials Science and Engineering, Jiangsu University, Zhenjiang 212013, China
3. School of Materials Science and Engineering, Northwestern Polytechnical University, Xi'an 710072, China
4. State Key Laboratory for Mechanical Behavior of Materials, Xi'an Jiaotong University, Xi'an 710049, China
5. Department of Chemistry, College of Science, King Saud University, Riyadh 11451, Saudi Arabia
* Correspondence: jyyangj@ujs.edu.cn (J.Y.); jhxu@just.edu.cn (J.X.)

Abstract: Cu$_3$SbSe$_4$ is a potential p-type thermoelectric material, distinguished by its earth-abundant, inexpensive, innocuous, and environmentally friendly components. Nonetheless, the thermoelectric performance is poor and remains subpar. Herein, the electrical and thermal transport properties of Cu$_3$SbSe$_4$ were synergistically optimized by S alloying. Firstly, S alloying widened the band gap, effectively alleviating the bipolar effect. Additionally, the substitution of S in the lattice significantly increased the carrier effective mass, leading to a large Seebeck coefficient of ~730 µVK^{-1}. Moreover, S alloying yielded point defect and Umklapp scattering to significantly depress the lattice thermal conductivity, and thus brought about an ultralow κ_{lat} ~0.50 Wm^{-1}K^{-1} at 673 K in the solid solution. Consequently, multiple effects induced by S alloying enhanced the thermoelectric performance of the Cu$_3$SbSe$_4$-Cu$_3$SbS$_4$ solid solution, resulting in a maximum ZT value of ~0.72 at 673 K for the Cu$_3$SbSe$_{2.8}$S$_{1.2}$ sample, which was ~44% higher than that of pristine Cu$_3$SbSe$_4$. This work offers direction on improving the comprehensive TE in solid solutions via elemental alloying.

Keywords: Cu$_3$SbSe$_4$-based materials; solid solutions; S alloying; point defect; thermoelectric properties

Citation: Zhao, L.; Han, H.; Lu, Z.; Yang, J.; Wu, X.; Ge, B.; Yu, L.; Shi, Z.; Karami, A.M.; Dong, S.; et al. Realizing the Ultralow Lattice Thermal Conductivity of Cu$_3$SbSe$_4$ Compound via Sulfur Alloying Effect. *Nanomaterials* **2023**, *13*, 2730. https://doi.org/10.3390/nano13192730

Academic Editor: Andreu Cabot

Received: 13 September 2023
Revised: 4 October 2023
Accepted: 6 October 2023
Published: 8 October 2023

Copyright: © 2023 by the authors. Licensee MDPI, Basel, Switzerland. This article is an open access article distributed under the terms and conditions of the Creative Commons Attribution (CC BY) license (https://creativecommons.org/licenses/by/4.0/).

1. Introduction

Thermoelectric (TE) technology has the capability to directly and reversibly convert heat into electricity, making it a promising source of clean energy. It plays a significant role in addressing the challenges posed by the energy and environmental crises [1–3]. Numerous TE materials are currently under exploration for power generation and solid-state cooling applications, leveraging the Seebeck and Peltier effects, respectively [4], such as skutterudites [5], half-Heusler compounds [6], Zintl phases [7], chalcogenides [8], oxides [9,10], and high-entropy alloys [11]. Commonly, the conversion efficiency of TE materials is assessed using the dimensionless figure of merit, $ZT = S^2\sigma T/\kappa$, where S, σ, T, and κ stand for the Seebeck coefficient, electrical conductivity, absolute temperature in Kelvin, and total thermal conductivity (comprising lattice part κ_{lat} and electronic part κ_{ele}), respectively [12,13]. Actually, achieving high conversion efficiency (η) necessitates a higher power factor (PF = $S^2\sigma$) and/or lower κ [14–20]. Unfortunately, it is difficult to simultaneously optimize the S, σ, and κ_{ele} in the given TE material due to their strong coupling effects [12,21]. Nevertheless, κ_{lat} stands as the sole independently regulated TE parameter, leading to extensive research over the last two decades [16,17,19].

Copper-based chalcogenides have garnered significant attention because of their relatively favorable electrical transport and low thermal transport properties [22–25]. In addition, thermoelectric minerals like germanites, colusites, tetrahedrites, and other materials also have rather high ZT values [26–28]. Among them, the Cu$_3$SbSe$_4$ compound is a p-type semiconductor, featuring a narrow band gap of ~0.29 eV [29,30]. More importantly, its components are earth-abundant, inexpensive, non-toxic, and environmentally friendly [31,32]. However, its high κ and low σ, stemming from low carrier concentration and mobility, present challenges that hinder its practical use. Extensive efforts have been implemented to enhance the TE performance of Cu_3SbSe_4, including elemental doping [33–37], band engineering [38–40], and nanostructure modification [41,42]. These approaches have potential in improving the carrier concentration of (n), S, or κ_{lat}, and thus leading to an appealing figure of merit. Although high n can enhance σ, it has a negative impact on S and result in an increase in κ_{ele}. The TE performance of Cu_3SbSe_4 falls significantly short of that of Cu-based chalcogenides due to these two inherent issues. On one side, the narrow energy band gap of ~0.29 eV leads to bipolar diffusion, causing deterioration in electrical properties [29,30]. On the other side, the high thermal conductivity (κ_{lat}) inherently arises from its composition comprising lightweight elements and a diamond-like structure [25]. In other words, optimizing carrier concentration alone proves challenging in further enhancing the TE performance.

The formation of a solid solution via elemental alloying is an effective strategy for depressing the κ_{lat} and thereby enhancing the TE performance. For example, Skoug et al. demonstrated that the substitution of Ge on Sn sites can lead to the formation of $Cu_2Sn_{1-x}Ge_xSe_3$ solid solutions, synergically optimizing the TE properties [43]. Jacob et al. reported that a high ZT_{max} value of ~0.42 was obtained in the $Cu_2Ge(S_{1-x}Se_x)_3$ system via Se alloying [44]. Wang et al. enhanced the TE properties of $Cu_2Ge(Se_{1-x}Te_x)_3$ by incorporating Te on the Se site, resulting in a ZT_{max} of ~0.55, which was 62% higher than that of the matrix [45]. The afore-mentioned research give us an idea that the $Cu_3Sb(Se_{1-x}S_x)_4$ solid solution is an effectively strategy for enhancing the thermoelectric performance of the Cu_3SbSe_4 compound via S alloying. Moreover, the development of TE materials with more cost-efficient constituent elements is of significant importance for large-scale practical applications.

Herein, we present the synthesis and thermoelectric characterization of the $Cu_3Sb(Se_{1-x}S_x)_4$ solid solutions with x covering the whole range from 0 to 1. The results demonstrate that the Cu_3SbSe_4-Cu_3SbS_4 solid solutions exhibit an extremely high Seebeck coefficient and ultralow thermal conductivity. Firstly, S alloying can widen the band gap, alleviating the bipolar effect. Additionally, S substitution in the lattice can significantly increase the carrier effective mass, leading to a remarkably high Seebeck coefficient of ~730 μVK^{-1}. Moreover, the κ_{lat} can be significantly depressed owing to point defect scattering and Umklapp scattering, thus obtaining a minimum κ_{lat} of ~0.50 $Wm^{-1}K^{-1}$. Consequently, the multiple effects of S alloying boost the TE performance of the Cu_3SbSe_4-Cu_3SbS_4 solid solution, and a maximum ZT value of ~0.72 at 673 K is obtained for the $Cu_3SbSe_{2.8}S_{1.2}$ sample.

2. Experimental Procedures

2.1. Synthesis

The $Cu_3Sb(Se_{1-x}S_x)_4$ solid solutions with varying S content (x = 0, 0.1, 0.2, 0.3, 0.4, 0.5, 0.6, 0.7, 0.8, and 1) were synthesized by vacuum melting and plasma-activated sintering (Ed-PAS III, Elenix Ltd., Zama, Japan). Concretely, the synthesis was divided into two steps. The first step was to synthesize the primary powders. Firstly, the starting materials, consisting of high-purity components (Cu: 99.99 wt.%; Sb: 99.99 wt.%; Se: 99.999 wt.%; S: 99.99 wt.%) corresponding to the nominal composition of $Cu_3Sb(Se_{1-x}S_x)_4$ (x = 0–1), were carefully sealed in the quartz tube under high vacuum conditions (<10^{-3} Pa). Afterwards, the sealed tubes were incrementally heated to 1173 K with a controlled rate of 20 K/h and maintained at 1173 K for a duration of 12 h. Following a holding period, the tubes were cooled down with a relatively low rate of 10 K/h until reaching 773 K, and finally the samples were quenched into water. Subsequently, the acquired quenched ingots underwent direct annealing at 573 K for a period of 48 h to facilitate the uniformity of chemical compositions. After this step, the obtained ingots were finely pulverized using an agate mortar to produce uniform powders. The second step was to synthesize the target samples. The resultant powders were then introduced into a graphite die of Ø12.7 mm in diameter and treated using the PAS technique at 673 K for a duration of 5 min while applying an axial pressure of 50 MPa. In detail, the sintering temperature reached to 523 K after an activation time of 10 s under the activation voltage of 20 V and the activation current of 300 A, and then the current was manually adjusted to increase by a rate of 1.5 K/s to reach the desired sintering temperature of 673 K after 225 s; the temperature was then held for 300 s. Ultimately, the samples were furnace-cooled to room temperature.

2.2. Characterization

The X-ray diffraction (XRD) patterns for the $Cu_3Sb(Se_{1-x}S_x)_4$ (x = 0–1) solid solutions were conducted using a Bruker D8 advance instrument, which was equipped with Cu Kα radiation (λ = 1.5418 Å). Lattice parameters were refined using the Rietveld method, employing the HighScore Plus computer program for analysis. The morphologies and compositions of the afore-mentioned solid solutions were performed by a Nova NanoSEM450 (FESEM) and a JEM-2010F (HRTEM), equipped with a detector of energy-dispersive X-ray spectroscopy (EDS).

2.3. Thermoelectric Property Measurements

The as-sintered cylinders were processed into bars of 10 mm × 2 mm × 2 mm and disks of Ø12.7 mm × 2 mm. The bars were used for concurrently measuring σ and S by the commercial measuring system (LINSEIS, LSR-3) under a helium atmosphere, spanning a temperature range from room temperature to 673 K. Thermal conductivity was calculated using the equation of $\kappa = DC_p\rho$. Herein, the D, C_p, and ρ stand for the thermal diffusivity, specific heat, and density, respectively. The disks were used for simultaneously measuring D and C_p by utilizing a Laser Flash apparatus of Netzsch (LFA-457) under a static argon atmosphere. The ρ of the $Cu_3Sb(Se_{1-x}S_x)_4$ (x = 0–1) solid solutions were conducted using Archimedes' methods. The relative densities, in relation to the theoretical density of 5.86 g cm^{-3}, have been provided in Table S1. The n (carrier concentration) and μ (carrier mobility) of the afore-mentioned solid solutions at 300 K were performed using the Hall effect system (LAKE SHORE, 7707 A) according to the van der Pauw method under a magnetic field strength of 0.68 T.

3. Results and discussion

3.1. Crystal Structure

The crystal structures and phase compositions for the $Cu_3Sb(Se_{1-x}S_x)_4$ (x = 0–1) samples were performed by XRD. Figure 1a shows the crystal structure of tetragonal Cu_3SbSe_4, with blue, gray, and green atoms representing Cu, Sb, and Se, respectively. As displayed in Figure 1b, the major diffraction peaks of the pristine sample (x = 0) are fully indexed to the zinc-blende-based tetragonal structure (I-42m space group) of Cu_3SbSe_4 (JCPDS No. 85-0003) without any detectable impurities [29]. With increasing S content (0 < x < 1), a continuous shift of the (112) diffraction peak towards higher angles can be seen (Figure 1c), demonstrating that S atoms replace Se at the Se site to form $Cu_3Sb(Se_{1-x}S_x)_4$ solid solutions. The shift in the diffraction peak can be ascribed to the smaller radius of S^{2-} (1.84 Å) in comparison to Se^{2-} (1.98 Å) [46]. For x = 1, the XRD peaks match the pattern of Cu_3SbS_4 (JCPDS No. 35-0581) [47].

The Rietveld refinement profiles of the $Cu_3Sb(Se_{1-x}S_x)_4$ (x = 0.3) samples based on the famatinite crystal structure are shown in Figure 1d. The data of the final agreement factors (R_p, R_{wp}, and R_{exp}) of $Cu_3Sb(Se_{1-x}S_x)_4$ (x = 0–1) samples are listed in Table S2. The lattice parameter exhibits a linear decrease with increasing S concentration, and closely follows the expected Vegard's law relationship [48] (Figure 1e), indicating the formation of Cu_3SbSe_4-Cu_3SbS_4 solid solutions.

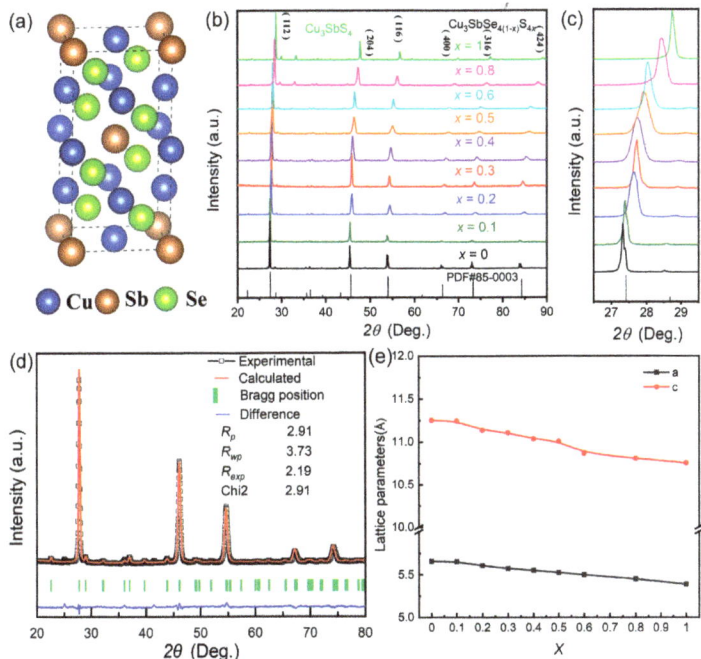

Figure 1. (a) The crystal structure of Cu_3SbSe_4; (b) X-ray diffraction (XRD) patterns and (c) magnified diffraction peaks corresponding to the (112) planes of $Cu_3Sb(Se_{1-x}S_x)_4$ (x = 0–1) samples; (d) Rietveld refinement profile of x = 0.3 solid solution; (e) Alterations in lattice parameters as S concentration varies.

3.2. Microstructure

The morphologies and chemical compositions of the $Cu_3Sb(Se_{1-x}S_x)_4$ (x = 0.3) sample were characterized by a SEM equipped with an EDS detector (Figure 2). As presented in Figure 2a,b, the SEM images of fracture surfaces (x = 0.3) indicated that they were isotropic materials. The nanopores (marked by the blue dotted circles) were observed on the fracture surface due to the Se/S volatilization of the synthesis process of the sample (Figure 2a), which can contribute to blocking the transport of mid-wavelength phonons [47]. To investigate the composition of the sample, we observed its polished surface (Figure 2c). According to the EDS elemental mapping (Figure 2d–h), the four constituent elements were uniformly distributed with no distinct micro-sized aggregations. This was combined with a back-scattered electron (BSE) image and elemental ratios (%), where Cu, Sb, Se, and S were present in proportions of 40.07:12.68:31.26:15.59 (as depicted in Figure S1), which demonstrated the formation of the Cu_3SbSe_4-Cu_3SbS_4 (x = 0.3) solid solution.

The morphologies and compositions of $Cu_3Sb(Se_{1-x}S_x)_4$ (x = 0.3) were further investigated at nanoscale using high-resolution TEM (HRTEM) (Figure 3). The TEM images demonstrated that many nanophases were distributed in the sample, and elemental mapping taking over the entire region revealed that the four constituent elements (Cu, Sb, Se, and S) were uniformly dispersed within the Cu_3SbSe_4-Cu_3SbS_4 solid solution (Figures 3a and S2). As presented in Figure 3b, the grain boundary (indicated by blue dot lines) could be clearly observed in the sample. Meanwhile, as shown in Figure 3b,c, the crossed fringes, with interplanar spacing of 3.26 Å and 1.99 Å corresponded to the (112) and (204) planes of Cu_3SbSe_4, respectively [49]. Additionally, the SAED pattern taken from the Figure 3c along the [110] zone axis is displayed in Figure 3d. The ordered diffraction spots can be indexed to the (002), ($1\bar{1}0$), and ($1\bar{1}2$) planes of Cu_3SbSe_4, whose interplanar spacings are 5.64 Å, 4.06 Å, and 3.26 Å, respectively [50].

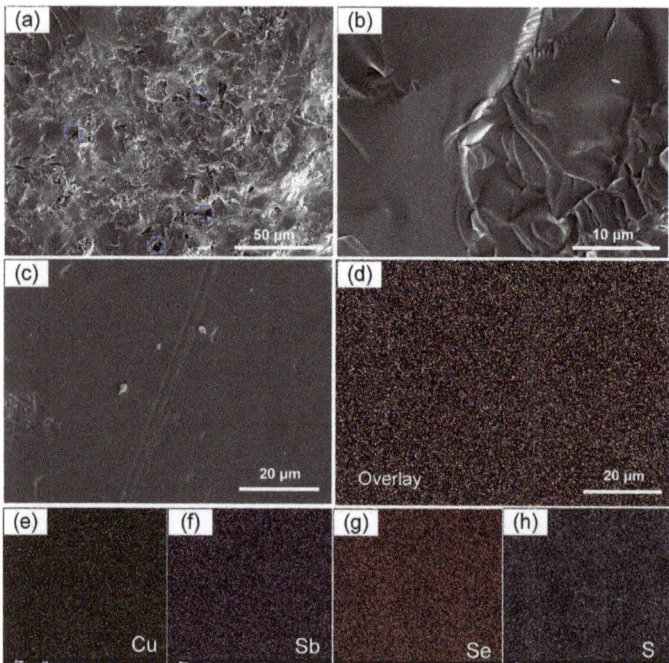

Figure 2. (**a**) SEM image of the fracture surfaces of the $Cu_3SbSe_{2.8}S_{1.2}$ sample; (**b**) high magnification images of (**a**); (**c**) the corresponding EDS mapping for all constituent elements of selected region in (**b**); (**c**) SEM images of the polished surfaces of the $Cu_3SbSe_{2.8}S_{1.2}$ sample; (**d**) The corresponding elemental mapping by EDS, obtained by overlaying the respective EDS signals directly arising from Cu (**e**), Sb (**f**), Se (**g**), and S (**h**).

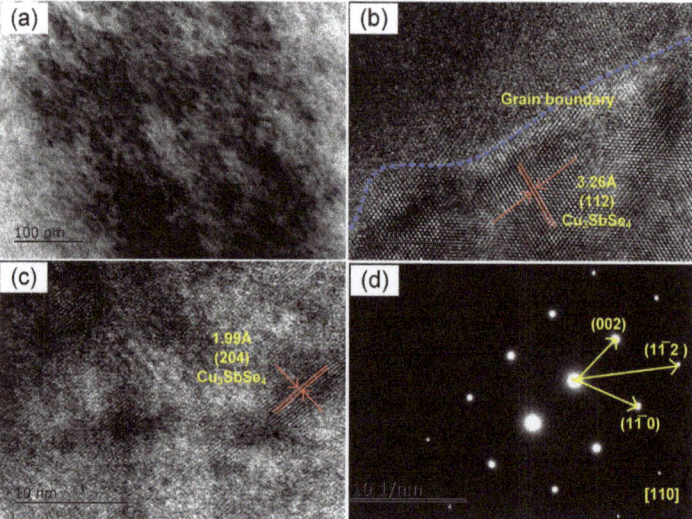

Figure 3. (**a**) The low-magnification image; (**b**,**c**) high-resolution TEM images; (**d**) SAED pattern taken from (**c**) of $Cu_3Sb(Se_{1-x}S_x)_4$ (x = 0.3) sample.

3.3. Charge Transport Properties

To explore the effects of S alloying on the TE properties of the $Cu_3Sb(Se_{1-x}S_x)_4$ (x = 0–1) solid solutions, the charge transport properties were conducted. The temperature dependence of electrical conductivity (σ) of the $Cu_3Sb(Se_{1-x}S_x)_4$ (x = 0–1) solid solutions is displayed in Figure 4a. The pristine Cu_3SbSe_4 exhibited a monotonous increase in σ with rising temperature, demonstrating characteristic behavior of a non-degenerate semiconductor. For the $x > 0.2$ samples, the samples showed a transition from non-degenerate semiconductors to a partially degenerate regime [51]. The σ exhibited an initial decrease followed by an increase, with the minimum value occurring at ~473 K, indicating its association with bipolar conduction [38,52]. The σ of S alloying samples increased with the S contents until x = 0.3, after which it started to decrease with a higher S content. Notably, the σ improved from ~4.6 S/cm of pristine Cu_3SbSe_4 to ~42 S/cm of x = 0.3 solid solution at room temperature, arising from the augmented carrier concentration (Table S1). It is worth noting that the solid solutions with high S content ($x > 0.5$) had lower σ compared to the pristine Cu_3SbSe_4, which was ascribed to the reduced n (carrier concentration) and diminished μ (carrier mobility). Furthermore, due to the intensified lattice vibration at elevated temperatures, the solid solutions exhibited lower σ than the pristine sample at high temperatures, indicating that the intensified lattice vibration in the solid solutions at elevated temperatures hindered the carrier migration [40,53].

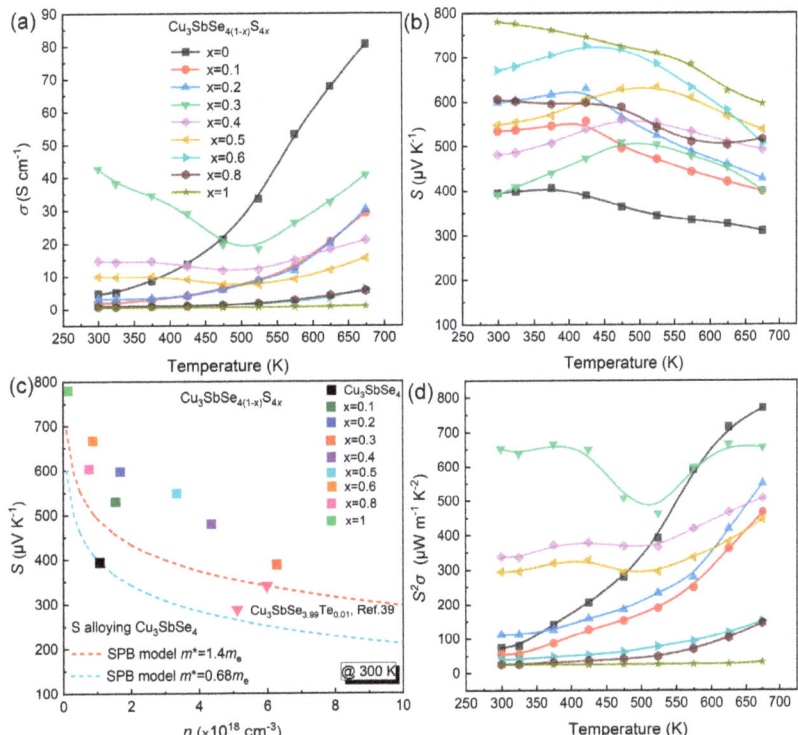

Figure 4. Temperature-dependent (**a**) electrical conductivity σ; (**b**) Seebeck coefficient S, the inset is E_g; (**c**) Pisarenko relationship with m^* in this work compared with other works at room temperature. The indigo and red broken line represent the Pisarenko relationship with $m^* \sim 0.68$ and $1.4~m_e$, respectively. (**d**) power factor $S^2\sigma$ of $Cu_3Sb(Se_{1-x}S_x)_4$ (x = 0–1) samples.

Figure 4b illustrates the temperature-dependent S of the $Cu_3Sb(Se_{1-x}S_x)_4$ (x = 0–1) samples. The p-type semiconductor behavior of solid solutions, characterized by dominant hole carriers, was evidenced by the positive S value observed across the entire temperature range.

Notably, the S value of the samples exhibited an initial ascent followed by a subsequent descent as the temperature rose, ultimately reaching its zenith at ~473 K. This behavior can be attributed to the influence of the bipolar effect [54]. The maximum S of ~730 μVK^{-1} was obtained from the x = 0.6 solid solution. We calculated the E_g of the $Cu_3Sb(Se_{1-x}S_x)_4$ (x = 0–1) samples with the formula: $E_g = 2eS_{max}T$, where E_g, e, S_{max}, and T represent the band gap, elementary charge, maximal Seebeck coefficient, and the associated temperature, respectively [55]. The calculated E_g for the pristine Cu_3SbSe_4 of ~0.30 eV aligned well with the reported literature [36,56]; the results are displayed in Figure S3. Consequently, the introduction of alloyed S played a role in enlarging E_g from ~0.30 eV to ~0.69 eV, thus widening the band gap to alleviate the bipolar effect. For the semiconductors, we note that the increase in S (|S|) was directly proportional to the carrier effective mass and $n^{-2/3}$. We calculated the Pisarenko relation between |S| and n (indigo and red dashed lines with m^* ~ 0.68 and 1.4 m_e, respectively) based on the single parabolic band model (SPB), as follows [57,58]:

$$S = \frac{8\pi^2 k_B^2}{3e\hbar^2} m * T \left(\frac{\pi}{3n}\right)^{2/3} \quad (1)$$

where k_B, $\hbar$ represent the Boltzmann constant, and Planck constant, respectively. The calculated m^* was significantly enhanced from 0.68 for pristine Cu_3SbSe_4 to 5.03 m_e for the x = 0.6 sample (Table S1). As seen in Figure 4c, the calculated m^* based on S (experimental values) of $Cu_3Sb(Se_{1-x}S_x)_4$ (x = 0.1–1) samples were above the Pisarenko line. Furthermore, the m^* depended directly on the E_g ($\frac{\hbar^2 k_B^2}{2m*} = E\left(1 + \frac{E}{E_g}\right)$), where E the energy of electron states), which deviated from a single Kane band model [21,59,60], thus confirming the large S was related to E_g and m^*. Consequently, the decreased carrier concentration ($x > 0.5$) and increased m^*, resulted in the significant enhancement of S.

The temperature dependence of the power factors ($S^2\sigma$) of the $Cu_3Sb(Se_{1-x}S_x)_4$ (x = 0–1) samples are presented in Figure 4d. The $S^2\sigma$ of Cu_3SbSe_4-Cu_3SbS_4 solid solutions exhibited a similar temperature-dependent behavior as the electrical conductivity (σ). The temperature-dependent trend observed in the $S^2\sigma$ was mirrored in the behavior of the σ for the Cu_3SbSe_4-Cu_3SbS_4 solid solutions. Owing to their relatively elevated σ and S values, these samples demonstrated larger $S^2\sigma$ values compared to the pristine Cu_3SbSe_4, particularly within the lower temperature range. Notably, the x = 0.3 sample achieved a larger $S^2\sigma$ value than the other samples, and the peak $S^2\sigma$ value of $Cu_3SbSe_{2.8}S_{1.2}$ sample was ~670 μW m^{-1} K^{-2} at 673 K.

3.4. Thermal Transport Properties

The temperature dependence of the thermal transport properties of the $Cu_3Sb(Se_{1-x}S_x)_4$ (x = 0–1) solid solutions are presented in Figure 5 and Figure S4. Obviously, the κ_{tot} decreased with increasing temperature, mainly attributed to the increased scattering by lattice vibrations at elevated temperatures [8,40,53] (Figure 5a). For instance, the κ_{tot} of pristine Cu_3SbSe_4 decreased from ~3.11 Wm^{-1}K^{-1} at 300 K to ~1.03 Wm^{-1}K^{-1} at 673 K. Similarly, the κ_{tot} of the x = 0.5 sample decreased from ~1.37 Wm^{-1}K^{-1} at 300 K to ~0.52 Wm^{-1}K^{-1} at 673 K. Generally, the κ_{lat} can be obtained by subtracting the electronic part (κ_{ele}) from the κ_{tot} using the Wiedeman–Franz relationship (the details are displayed in Supplementary Material) [61,62]:

$$\kappa_{ele} = L\sigma T \quad (2)$$

where L is the Lorenz number and it can be expressed as Equation (3) [63,64]:

$$L = 1.5 + \exp\left[\frac{-|S|}{116}\right] \quad (3)$$

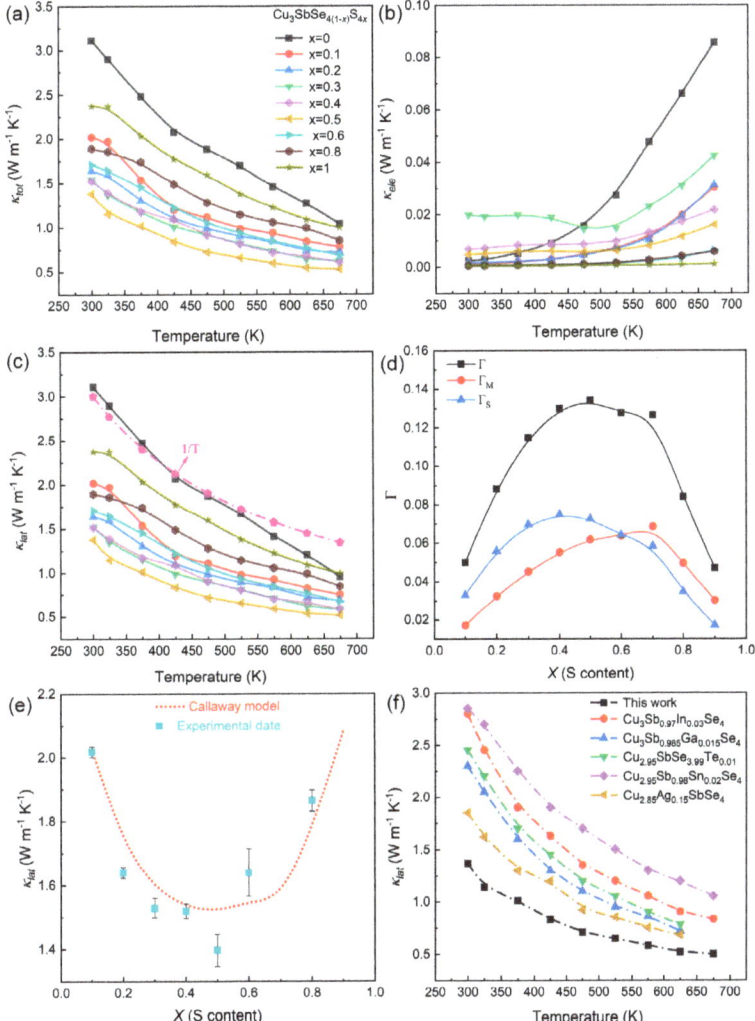

Figure 5. Temperature-dependent (**a**) total thermal conductivity κ_{tot}; (**b**) electronic thermal conductivity κ_{ele}; (**c**) lattice thermal conductivity κ_{lat} of $Cu_3Sb(Se_{1-x}S_x)_4$ ($x = 0$–1) samples, and (**d**) imperfection scaling parameters; (**e**) κ_{lat} at 300 K, the red dotted lines is calculated by the Callaway model; (**f**) the comparison of κ_{lat} of Cu_3SbSe_4-based materials [33,34,36,38,39].

The calculated L values of the $Cu_3Sb(Se_{1-x}S_x)_4$ ($x = 0$–1) samples ranged from 1.5 to 1.6 W Ω K^{-2}, and the results are listed in Figure S5. Owing to the enhanced carrier concentration ($x < 0.6$), the κ_{ele} showed a slight increase at low temperature after S alloying, as described in Figure 5b. The κ_{lat} of the $Cu_3Sb(Se_{1-x}S_x)_4$ ($x = 0$–1) samples is plotted in Figure 5c, indicating a significant decrease within the measured temperature range after S alloying.

To explore the effects of S alloying on the phonon scattering and the significant reduction in κ_{lat}, the κ_{lat} of the $Cu_3Sb(Se_{1-x}S_x)_4$ ($x = 0$–1) compounds was evaluated at room temperature by the Debye–Callaway model. The primary scattering mechanisms under consideration were point defect scattering and Umklapp scattering. Then, the κ_{lat} of the pristine ($\kappa_{lat}^{pristine}$) and S-alloyed (κ_{lat}) Cu$_3$SbSe$_4$ compounds could be computed based on the Debye–Callaway model [48,65,66]:

$$\frac{\kappa_{lat}}{\kappa_{lat}^{pristine}} = \frac{\arctan(u)}{u}, \quad u^2 = \frac{\pi^2 \theta_D \Omega}{hv^2} \kappa_{lat}^{pristine} \Gamma \quad (4)$$

where u, θ_D, Ω, h, and v represent the scaling parameter, Debye temperature, volume per atom, Planck constant, and average speed of sound, respectively (herein, θ_D = 131 K and v = 1991.2 m/s [46]). Γ is the imperfection scale parameter, which is associated with the Γ_m (mass fluctuation) and Γ_s (strain field fluctuation) [67]:

$$\Gamma = \Gamma_m + \Gamma_s = x(1-x)\left[\left(\frac{\Delta M}{M}\right)^2 + \varepsilon\left(\frac{\Delta r}{r}\right)^2\right] \quad (5)$$

where x, $\Delta M/M$ and $\Delta r/r$ are the S concentration in one molecular, the relative change of atomic mass, and atomic radius owing to the replacement of Se with S, respectively. The ε value can be computed using the following formula [68]:

$$\varepsilon = \frac{2}{9}\left(\frac{6.4\gamma(1+v_p)}{1-v_p}\right)^2 \quad (6)$$

where, γ and v_p are the Grüneisen parameter and Poisson ratio, respectively (here, γ = 1.3 [46] and v_p = 0.35 [69]).

The values of Γ_m and Γ_s for the Cu$_3$Sb(Se$_{1-x}$S$_x$)$_4$ compounds are presented in Table S3. Figure 5d shows how the scattering parameters Γ_m and Γ_s changed with varying Se-alloying levels. It was observed that Γ_m was smaller than Γ_s when the S content $x \leq 0.6$, indicating that the Γ_s (strain field fluctuation) contributed greatly to the drop of κ_{lat}. As for the $x > 0.6$ samples, the Γ_m (mass fluctuation) was the dominant. It is commonly accepted that the atomic radius of S is different from that of the Se atom, inducing a localized lattice distortion and leading to local field fluctuations that hinder the propagation of heat-carrying phonons [70,71]. However, with increasing S content, the mass fluctuation gradually became the dominant factor. The experimental κ_{lat} closely aligned with the curve calculated by the Callaway model (Figure 5e), suggesting that point defects made a great contribution to suppress the κ_{lat} in the Cu$_3$Sb(Se$_{1-x}$S$_x$)$_4$ solid solutions [48]. For a more comprehensive evaluation of our results, a comparison of the our κ_{lat} data with the recently reported values of Cu$_3$SbSe$_4$ are illustrated in Figure 5f [33,34,36,38,39]. Remarkably, the Cu$_3$Sb(Se$_{1-x}$S$_x$)$_4$ (x = 0.5) sample achieved an outstandingly low κ_{lat} of ~0.50 W m^{-1} K^{-1} at 673 K.

3.5. Figure of Merit (ZT)

The temperature-dependent ZT of the Cu$_3$Sb(Se$_{1-x}$S$_x$)$_4$ (x = 0–1) samples are illustrated in Figure 6a. With the benefit of the collaborative enhancement of electrical and thermal transport properties, the x = 0.3 sample attained a maximum ZT value of ~0.72 at 673 K, which was 44% higher than that of pristine Cu$_3$SbSe$_4$. To further analyze our TE properties, the comparison of the ZT$_{max}$ of the Cu$_3$SbSe$_4$-based materials is given in Figure 6b [33–40,72]. Obviously, our ZT$_{max}$ of 0.72 was higher than that of other Cu$_3$SbSe$_4$-based materials, such as Cu$_3$Sb$_{0.97}$In$_{0.03}$Se$_4$ ~0.5, Cu$_3$Sb$_{0.985}$Ga$_{0.015}$Se$_4$ ~0.54, Cu$_3$SbSe$_{3.99}$Te$_{0.01}$ ~0.62, Cu$_3$Sb$_{0.92}$Sn$_{0.08}$S$_{3.75}$Se$_{0.25}$ ~0.67, Cu$_{2.95}$Sb$_{0.96}$Ge$_{0.04}$Se$_4$ ~0.70, and Cu$_{2.95}$Sb$_{0.98}$Sn$_{0.02}$Se$_4$ ~0.7 and is comparable to the ZT values for Cu$_3$Sb$_{0.98}$Bi$_{0.02}$Se$_{3.99}$Te$_{0.01}$ ~0.76 and Cu$_3$Sb$_{0.91}$Sn$_{0.03}$Hf$_{0.06}$Se$_4$ ~0.76. Although the Cu$_3$Sb(Se$_{1-x}$S$_x$)$_4$ (x = 0–1) samples had relative low ZT values in comparison with other high-performance TE materials, further enhancements of the ZT values can potentially be achieved by tuning the carrier concentration, dual-incorporation, and/or introducing band engineering.

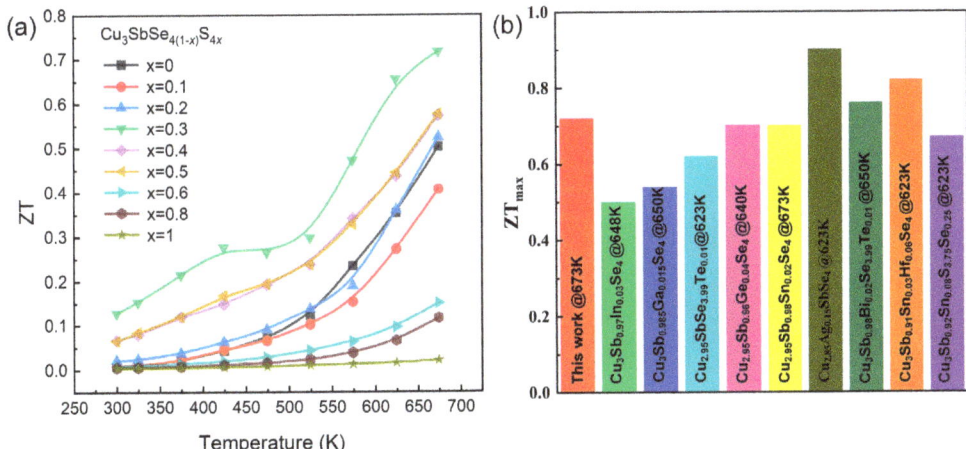

Figure 6. (a) Temperature-dependent figure of merit (ZT); (b) Comparison of ZT_{max} of Cu_3SbSe_4-based materials [33–40,72].

4. Conclusions

In summary, a series of Cu_3SbSe_4-Cu_3SbS_4 solid solutions were synthesized by vacuum melting and plasma-activated sintering (PAS) techniques, and the effects of S alloying on TE performance were investigated. S alloying can widen the band gap, effectively alleviating the bipolar effect. Additionally, the S (Seebeck coefficient) was significantly improved because of the increased m^*. Furthermore, the substitution of S for Se in Cu_3SbSe_4 lattice led to noticeable local distortions, yielding large strain and mass fluctuations to suppress the κ_{lat}, thus decreasing the κ_{lat} and κ_{tot} to ~0.50 $Wm^{-1}K^{-1}$ and ~0.52 $Wm^{-1}K^{-1}$ at 673 K, respectively. Consequently, a peak ZT value of ~0.72 was obtained at 673 K for the $Cu_3Sb(Se_{1-x}S_x)_4$ ($x = 0.3$) sample. Based on these results, it is speculated that further improvement in the figure of merit of $Cu_3Sb(Se_{1-x}S_x)_4$ solid solutions can be obtained by enhanced electrical transport properties. Our research offers a new strategy to develop high-performance TE materials in solid solutions via elemental alloying.

Supplementary Materials: The supporting information can be downloaded at: https://www.mdpi.com/article/10.3390/nano13192730/s1.

Author Contributions: Conceptualization, L.Z., J.Y., B.G., L.Y., Z.S., A.M.K., S.D., S.H., G.Q. and J.X.; Methodology, L.Z., H.H., J.Y., X.W., B.G., G.Q. and J.X.; Validation, Z.L. and S.H.; Formal analysis, L.Z., J.Y., X.W., B.G., L.Y., Z.S., A.M.K., S.D. and G.Q.; Investigation, L.Z., H.H., Z.L. and X.W.; Resources, J.Y. and J.X.; Data curation, B.G., Z.S., A.M.K. and S.H.; Writing—original draft, L.Z., H.H. and Z.L.; Writing—review & editing, J.Y.; Visualization, S.D.; Supervision, L.Y., G.Q. and J.X.; Project administration, S.H.; Funding acquisition, L.Y., G.Q. and J.X. All authors have read and agreed to the published version of the manuscript.

Funding: This work was funded by the National Natural Science Foundation of China (51572111, 52071159, 52172090), the Natural Science Foundation (BK20210779), the University-Industry Research Cooperation Project (BY20221151), and the Universities Natural Science Research Project (21KJB430019) of Jiangsu Province. This work was also funded by the Researchers Supporting Project Number (RSPD2023R764), King Saud University, Riyadh, Saudi Arabia.

Data Availability Statement: Data will be made available on reasonable request.

Conflicts of Interest: The authors declare that they have no known competing financial interests or personal relationships that could have appeared to influence the work reported in this paper.

References

1. Bell, L.E. Cooling, heating, generating power, and recovering waste heat with thermoelectric systems. *Science* **2008**, *321*, 1457–1461. [CrossRef]
2. Snyder, G.J.; Toberer, E.S. Complex thermoelectric materials. *Nat. Mater.* **2011**, *473*, 66–69.
3. Shi, X.L.; Zou, J.; Chen, Z.-G. Advanced thermoelectric design: From materials and structures to devices. *Chem. Rev.* **2020**, *120*, 7399–7515.
4. DiSalvo, F.J. Termoelectric cooling and power generation. *Science* **1999**, *285*, 703–706. [CrossRef] [PubMed]
5. Tang, Y.L.; Gibbs, Z.M.; Agapito, L.A.; Li, G.D.; Kim, H.S.; Nardelli, M.B.; Curtarolo, S.; Snyder, G.J. Convergence of multi-valley bands as the electronic origin of high thermoelectric performance in $CoSb_3$ skutterudites. *Nat. Mater.* **2015**, *14*, 1223–1228. [PubMed]
6. Fu, C.G.; Bai, S.Q.; Liu, Y.T.; Tang, Y.S.; Chen, L.D.; Zhao, X.B.; Zhu, T.J. Realizing high figure of merit in heavy-band p-type half-Heusler thermoelectric materials. *Nat. Commun.* **2015**, *6*, 8144. [PubMed]
7. Song, J.; Song, H.Y.; Wang, Z.; Lee, S.; Hwang, J.Y.; Lee, S.Y.; Lee, J.; Kim, D.; Lee, K.H.; Kim, Y.; et al. Creation of two-dimensional layered Zintl phase by dimensional manipulation of crystal structure. *Sci. Adv.* **2019**, *5*, eaax0390.
8. Zhou, C.J.; Lee, Y.K.; Yu, Y.; Byun, S.J.; Luo, Z.Z.; Lee, H.; Ge, B.Z.; Lee, Y.L.; Chen, X.Q.; Lee, J.Y.; et al. Polycrystalline SnSe with a thermoelectric figure of merit greater than the single crystal. *Nat. Mater.* **2021**, *20*, 1378–1384.
9. Zhang, L.; Hong, X.; Chen, Z.Q.; Xiong, D.K.; Bai, J.M. Effects of In_2O_3 nanoparticles addition on microstructures and thermoelectric properties of $Ca_3Co_4O_9$ compounds. *Ceram. Int.* **2020**, *46*, 17763–17766. [CrossRef]
10. Dong, S.-T.; Yu, M.-C.; Fu, Z.; Lv, Y.-Y.; Yao, S.-H.; Chen, Y.B. High thermoelectric performance of NaF-doped $Bi_2Ca_2Co_2O_y$ ceramic samples. *J. Mater. Res. Technol.* **2022**, *17*, 1598–1604. [CrossRef]
11. Jiang, B.B.; Yu, Y.; Cui, J.; Liu, X.X.; Xie, L.; Liao, J.C.; Zhang, Q.H.; Huang, Y.; Ning, S.C.; Jia, B.H.; et al. High-entropy-stabilized chalcogenides with high thermoelectric performance. *Science* **2021**, *371*, 830–834. [CrossRef] [PubMed]
12. He, J.; Tritt, T.M. Advances in thermoelectric materials research: Looking back and moving forward. *Science* **2017**, *357*, 1369.
13. Yang, X.; Wang, C.Y.; Lu, R.; Shen, Y.N.; Zhao, H.B.; Li, J.; Li, R.Y.; Zhang, L.X.; Chen, H.S.; Zhang, T.; et al. Progress in measurement of thermoelectric properties of micro/nano thermoelectric materials: A critical review. *Nano Energy* **2022**, *101*, 107553.
14. Lan, R.; Otoo, S.L.; Yuan, P.Y.; Wang, P.F.; Yuan, Y.Y.; Jiang, X.B. Thermoelectric properties of Sn doped GeTe thin films. *Appl. Surf. Sci.* **2020**, *507*, 145025.
15. Pei, Y.Z.; Shi, X.Y.; LaLonde, A.; Wang, H.; Chen, L.D.; Snyder, G.J. Convergence of electronic bands for high performance bulk thermoelectrics. *Nature* **2011**, *473*, 66–69. [PubMed]
16. Zhang, Q.; Song, Q.C.; Wang, X.Y.; Sun, J.Y.; Zhu, Q.; Dahal, K.; Lin, X.; Cao, F.; Zhou, J.W.; Chen, S.; et al. Deep defect level engineering: A strategy of optimizing the carrier concentration for high thermoelectric performance. *Energy Environ. Sci.* **2018**, *11*, 933–940. [CrossRef]
17. Cheng, C.; Zhao, L.D. Anharmoncity and low thermal conductivity in thermoelectrics. *Mater. Today Phys.* **2018**, *4*, 50–57.
18. Zhao, H.B.; Yang, X.; Wang, C.Y.; Lu, R.; Zhang, T.; Chen, H.S.; Zheng, X.H. Progress in thermal rectification due to heat conduction in micro/nano solids. *Mater. Today Phys.* **2023**, *30*, 100941.
19. Biswas, K.; He, J.Q.; Blum, I.D.; Wu, C.I.; Hogan, T.P.; Seidman, D.N.; Dravid, V.P.; Kanatzidis, M.G. High-performance bulk thermoelectrics with all-scale hierarchical architectures. *Nature* **2012**, *489*, 414–418. [CrossRef]
20. Kim, S.; Lee, K.H.; Mun, H.A.; Kim, H.S.; Hwang, S.W.; Roh, J.W.; Yang, D.J.; Shin, W.H.; Li, X.S.; Lee, Y.H.; et al. Dense dislocation arrays embedded in grain boundaries for high-performance bulk thermoelectrics. *Science* **2015**, *348*, 109–114. [CrossRef]
21. Tan, G.J.; Zhao, L.D.; Kanatzidis, M.G. Rationally designing high-performance bulk thermoelectric materials. *Chem. Rev.* **2016**, *116*, 12123–12149.
22. Liu, H.L.; Shi, X.; Xu, F.F.; Zhang, L.L.; Zhang, W.Q.; Chen, L.D.; Li, Q.; Uher, C.; Day, T.; Snyder, G.J. Copper ion liquid-like thermoelectrics. *Nat. Mater.* **2012**, *11*, 422–425. [CrossRef]
23. Qiu, P.F.; Shi, X.; Chen, L.D. Cu-based thermoelectric materials. *Energy Storage Mater.* **2016**, *3*, 85–97.
24. Bo, L.; Li, F.J.; Hou, Y.B.; Zuo, M.; Zhao, D.G. Enhanced thermoelectric performance of Cu_2Se via nanostructure and compositional gradient. *Nanomaterials* **2022**, *12*, 640. [CrossRef] [PubMed]
25. Ge, B.Z.; Lee, H.; Zhou, C.J.; Lu, W.Q.; Hu, J.B.; Yang, J.; Cho, S.-P.; Qiao, G.; Shi, Z.; Chung, I. Exceptionally low thermal conductivity realized in the chalcopyrite $CuFeS_2$ via atomic-level lattice engineering. *Nano Energy* **2022**, *94*, 106941.
26. Kumar, V.P.; Paradis-Fortin, L.; Lemoine, P.; Caër, G.L.; Malaman, B.; Boullay, P.; Raveau, B.; Guélou, G.; Guilmeau, E. Crossover from germanite to renierite-type structures in $Cu_{22-x}Zn_xFe_8Ge_4S_{32}$ thermoelectric sulfides. *ACS Appl. Energy Mater.* **2019**, *2*, 7679–7689. [CrossRef]
27. Hagiwara, T.; Suekuni, K.; Lemoine, P.; Supka, A.R.; Chetty, R.; Guilmeau, E.; Raveau, B.; Fornari, M.; Ohta, M.; Orabi, R.R.; et al. Key role of d^0 and d^{10} cations for the design of semiconducting colusites: Large thermoelectric ZT in $Cu_{26}Ti_2Sb_6S_{32}$ compounds. *Chem. Mater.* **2021**, *33*, 3449–3456. [CrossRef]
28. Guélou, G.; Lemoine, P.; Raveau, B.; Guilmeau, E. Recent developments in high-performance thermoelectric sulphides: An overview of the promising synthetic colusites. *J. Mater. Chem.* **2021**, *9*, 773–795.
29. Yang, C.Y.; Huang, F.Q.; Wu, L.M.; Xu, K. New stannite-like p-type thermoelectric material Cu_3SbSe_4. *J. Phys. D Appl. Phys.* **2011**, *44*, 295404. [CrossRef]

30. Do, D.T.; Mahanti, S.D. Theoretical study of defects Cu_3SbSe_4: Search for optimum dopants for enhancing thermoelectric properties. *J. Alloys Compd.* **2015**, *625*, 346–354. [CrossRef]
31. Huang, Y.L.; Zhang, B.; Li, J.W.; Zhou, Z.Z.; Zheng, S.K.; Li, N.H.; Wang, G.W.; Zhang, D.; Zhang, D.L.; Han, G.; et al. Unconventional doping effect leads to ultrahigh average thermoelectric power factor in Cu_3SbSe_4-Based composites. *Adv. Mater.* **2022**, *34*, 2109952. [CrossRef] [PubMed]
32. García, G.; Palacios, P.; Cabot, A.; Wahnón, P. Thermoelectric properties of doped-Cu_3SbSe_4 compounds: A first-principles insight. *Inorg. Chem.* **2018**, *57*, 7321–7333. [CrossRef]
33. Zhang, D.; Yang, J.Y.; Jiang, Q.H.; Fu, L.W.; Xiao, Y.; Luo, Y.B.; Zhou, Z.W. Improvement of thermoelectric properties of Cu_3SbSe_4 compound by in doping. *Mater. Des.* **2016**, *98*, 150–154. [CrossRef]
34. Zhao, D.G.; Wu, D.; Bo, L. Enhanced thermoelectric properties of Cu_3SbSe_4 compounds via Gallium doping. *Energies* **2017**, *10*, 1524. [CrossRef]
35. Chang, C.H.; Chen, C.L.; Chiu, W.T.; Chen, Y.Y. Enhanced thermoelectric properties of Cu_3SbSe_4 by germanium doping. *Mater. Lett.* **2017**, *186*, 227–230. [CrossRef]
36. Wei, T.R.; Wang, W.H.; Gibbs, Z.M.; Wu, C.F.; Snyder, G.J.; Li, J.-F. Thermoelectric properties of Sn-doped p-type Cu_3SbSe_4: A compound with large effective mass and small band gap. *J. Mater. Chem. A* **2014**, *2*, 13527–13533. [CrossRef]
37. Kumar, A.; Dhama, P.; Banerji, P. Enhanced thermoelectric properties in Bi and Te doped p-type Cu_3SbSe_4 compound. In Proceedings of the DAE Solid State Physics Symposium 2017, Mumbai, India, 26–30 December 2017; Volume 1942, p. 140080.
38. Zhang, D.; Yang, J.Y.; Bai, H.C.; Luo, Y.B.; Wang, B.; Hou, S.H.; Li, Z.L.; Wang, S.F. Significant average ZT enhancement in Cu_3SbSe_4-based thermoelectric material via softening p-d hybridization. *J. Mater. Chem. A* **2019**, *7*, 17655–17656. [CrossRef]
39. Zhang, D.; Yang, J.Y.; Jiang, Q.H.; Zhou, Z.W.; Li, X.; Ren, Y.Y.; Xin, J.W.; Basit, A.; He, X.; Chu, W.J.; et al. Simultaneous optimization of the overall thermoelectric properties of Cu_3SbSe_4 by band engineering and phonon blocking. *J. Alloys Compd.* **2017**, *724*, 597–602. [CrossRef]
40. Wang, B.Y.; Zheng, S.Q.; Wang, Q.; Li, Z.L.; Li, J.; Zhang, Z.P.; Wu, Y.; Zhu, B.S.; Wang, S.Y.; Chen, Y.X.; et al. Synergistic modulation of power factor and thermal conductivity in Cu_3SbSe_4 towards high thermoelectric performance. *Nano Energy* **2020**, *71*, 104658. [CrossRef]
41. Xie, D.D.; Zhang, B.; Zhang, A.J.; Chen, Y.J.; Yan, Y.C.; Yang, H.Q.; Wang, G.W.; Wang, G.Y.; Han, X.D.; Han, G.; et al. High thermoelectric performance of Cu_3SbSe_4 nanocrystals with $Cu_{2-x}Se$ in situ inclusions synthesized by a microwave-assisted solvothermal method. *Nanoscale* **2018**, *10*, 14546–14553. [CrossRef] [PubMed]
42. Zhao, L.J.; Yu, L.H.; Yang, J.; Wang, M.Y.; Shao, H.C.; Wang, J.L.; Shi, Z.Q.; Wan, N.; Hussain, S.; Qiao, G.J.; et al. Enhancing thermoelectric and mechanical properties of p-type Cu_3SbSe_4-based materials via embedding Nanoscale Sb_2Se_3. *Mater. Chem. Phys.* **2022**, *292*, 126669.
43. Skoug, E.J.; Cain, J.D.; Morelli, D.T. Thermoelectric properties of the Cu_2SnSe_3–Cu_2GeSe_3 solid solution. *J. Alloys Compd.* **2010**, *506*, 18–21. [CrossRef]
44. Jacob, S.; Delatouche, B.; Péré, D.; Jacoba, A.; Chmielowski, R. Insights into the thermoelectric properties of the $Cu_2Ge(S_{1-x}Se_x)_3$ solid solutions. *Mater. Today Proc.* **2017**, *4*, 12349–12359.
45. Wang, R.F.; Dai, L.; Yan, Y.C.; Peng, K.L.; Lu, X.; Zhou, X.Y.; Wang, G.Y. Complex alloying effect on thermoelectric transport properties of $Cu_2Ge(Se_{1-x}Te_x)_3$. *Chin. Phys. B* **2018**, *27*, 067201.
46. Skoug, E.J.; Cain, J.D.; Morelli, D.T. High thermoelectric figure of merit in the Cu_3SbSe_4-Cu_3SbS_4 solid solution. *Appl. Phys. Lett.* **2011**, *98*, 261911. [CrossRef]
47. Lu, B.B.; Wang, M.Y.; Yang, J.; Hou, H.G.; Zhang, X.Z.; Shi, Z.Q.; Liu, J.L.; Qiao, G.J.; Liu, G.W. Dense twin and domain boundaries lead to high thermoelectric performance in Sn-doped Cu_3SbS_4. *Appl. Phys. Lett.* **2022**, *120*, 173901.
48. Zhao, K.P.; Blichfeld, A.B.; Eikeland, E.; Qiu, P.F.; Ren, D.D.; Iversen, B.B.; Shi, X.; Chen, L.D. Extremely low thermal conductivity and high thermoelectric performance in liquid-like $Cu_2Se_{1-x}S_x$ polymorphic materials. *J. Mater. Chem. A* **2017**, *5*, 18148–18156.
49. Li, J.M.; Ming, H.W.; Song, C.J.; Wang, L.; Xin, H.X.; Gu, Y.J.; Zhang, J.; Qin, X.Y.; Li, D. Synergetic modulation of power factor and thermal conductivity for Cu_3SbSe_4-based system. *Mater. Today Energy* **2020**, *18*, 100491.
50. Zhou, T.; Wang, L.J.; Zheng, S.Q.; Hong, M.; Fang, T.; Bai, P.P.; Chang, S.Y.; Cui, W.L.; Shi, X.L.; Zhao, H.Z.; et al. Self-assembled 3D flower-like hierarchical Ti-doped Cu_3SbSe_4 microspheres with ultralow thermal conductivity and high zT. *Nano Energy* **2018**, *49*, 221–229. [CrossRef]
51. Li, X.Y.; Li, D.; Xin, H.X.; Zhang, J.; Song, C.J.; Qin, X.Y. Effects of bismuth doping the thermoelectric properties of Cu_3SbSe_4 at moderate temperatures. *J. Alloys Compd.* **2013**, *561*, 105–108. [CrossRef]
52. Bhardwaj, R.; Bhattacharya, A.; Tyagi, K.; Gahtori, B.; Chauhan, N.S.; Bathula, S.; Auluck, S.; Dhar, A. Tin doped Cu_3SbSe_4: A stable thermoelectric analogue for the mid-temperature applications. *Mater. Res. Bull.* **2019**, *113*, 38–44.
53. Yang, J.; Zhang, X.Z.; Liu, G.W.; Zhao, L.J.; Liu, J.L.; Shi, Z.Q.; Ding, J.N.; Qiao, G.J. Multiscale structure and band configuration tuning to achieve high thermoelectric properties in n-type PbS bulks. *Nano Energy* **2020**, *74*, 104826.
54. Yang, X.X.; Gu, Y.Y.; Li, Y.P.; Guo, K.; Zhang, J.Y.; Zhao, J.T. The equivalent and aliovalent dopants boosting the thermoelectric properties of $YbMg_2Sb_2$. *Sci. China Mater.* **2020**, *63*, 437–443.
55. Goldsmid, H.J.; Sharp, J.W. Estimation of the thermal band gap of a semiconductor from Seebeck measurements. *J. Electron. Mater.* **1999**, *28*, 869–872. [CrossRef]

56. Skoug, E.J.; Cain, J.D.; Majsztrik, P.; Kirkham, M.; LaraCurzio, E.; Morelli, D.T. Doping effects on the thermoelectric properties of Cu_3SbSe_4. *Sci. Adv. Mater.* **2011**, *3*, 602–606. [CrossRef]
57. Cutler, M.; Leavy, J.F.; Fitzpatrick, R.L. Electronic transport in semimetallic cerium sulfide. *Phys. Rev.* **1964**, *133*, A1143–A1152.
58. Wei, T.-R.; Tan, G.; Zhang, X.; Wu, C.-F.; Li, J.-F.; Dravid, V.P.; Snyder, G.J.; Kanatzidis, M.G. Distinct impact of Alkali-Ion doping on electrical transport properties of thermoelectric p-Type polycrystalline SnSe. *J. Am. Chem. Soc.* **2016**, *138*, 8875–8882. [CrossRef] [PubMed]
59. Ge, B.Z.; Lee, H.; Im, J.; Choi, Y.; Kim, Y.-S.; Lee, J.Y.; Cho, S.-P.; Sung, Y.-E.; Choi, K.-Y.; Zhou, C.J.; et al. Engineering atomic-level crystal lattice and electronic band structure for extraordinarily high average thermoelectric figure of merit in n-type PbSe. *Energy Environ. Sci.* **2023**, *16*, 3994–4008. [CrossRef]
60. Zhou, C.J.; Yu, Y.; Lee, Y.L.; Ge, B.Z.; Lu, W.Q.; Miredin, O.C.; Im, J.; Cho, S.P.; Wuttig, M.; Shi, Z.Q.; et al. Exceptionally high average power factor and thermoelectric figure of merit in n-type PbSe by the dual incorporation of Cu and Te. *J. Am. Chem. Soc.* **2020**, *142*, 15172–15186. [CrossRef]
61. Mahan, G.D.; Bartkowiak, M. Wiedemann–Franz law at boundaries. *Appl. Phys. Lett.* **1999**, *74*, 953–954. [CrossRef]
62. Fu, Z.; Jiang, J.L.; Dong, S.-T.; Yu, M.-C.; Zhao, L.J.; Wang, L.; Yao, S.-H. Effects of Zr substitution on structure and thermoelectric properties of Bi_2O_2Se. *J. Mater. Res. Technol.* **2022**, *21*, 640–647.
63. Kim, H.S.; Gibbs, Z.M.; Tang, Y.G.; Wang, H.; Snyder, G.J. Characterization of Lorenz number with Seebeck coefficient measurement. *APL Mater.* **2015**, *3*, 041506.
64. Wang, B.Y.; Zheng, S.Q.; Chen, Y.X.; Wu, Y.; Li, J.; Ji, Z.; Mu, Y.N.; Wei, Z.B.; Liang, Q.; Liang, J.X. Band Engineering for Realizing Large Effective Mass in Cu_3SbSe_4 by Sn/La Co-doping. *J. Phys. Chem. C* **2020**, *124*, 10336–10343.
65. Callaway, J.; von Baeyer, H.C. Effect of point imperfections on lattice thermal conductivity. *Phys. Rev.* **1960**, *120*, 1149–1154.
66. Yang, J.; Meisner, G.P.; Chen, L.D. Strain field fluctuation effects on lattice thermal conductivity of ZrNiSn-based thermoelectric compounds. *Appl. Phys. Lett.* **2004**, *85*, 1140–1142.
67. Xie, H.Y.; Su, X.L.; Zheng, G.; Zhu, T.; Yin, K.; Yan, Y.G.; Uher, C.; Kanatzidis, M.G.; Tang, X.F. The role of Zn in chalcopyrite $CuFeS_2$: Enhanced thermoelectric properties of $Cu_{1-x}Zn_xFeS_2$ with in situ nanoprecipitates. *Adv. Energy Mater.* **2017**, *7*, 1601299.
68. Wan, C.L.; Pan, W.; Xu, Q.; Qin, Y.X.; Wang, J.D.; Qu, Z.X.; Fang, M.H. Effect of point defects on the thermal transport properties of $(La_xGd_{1-x})_2Zr_2O_7$: Experiment and theoretical model. *Phys. Rev. B* **2006**, *74*, 144109.
69. Xu, B.; Zhang, X.D.; Su, Y.Z.; Zhang, J.; Wang, Y.S.; Yi, L. Elastic Anisotropy and Anisotropic Transport Properties of Cu_3SbSe_4 and Cu_3SbS_4. *J. Phys. Soc. Jpn.* **2014**, *83*, 094606.
70. Yang, J.; Song, R.F.; Zhao, L.J.; Zhang, X.Z.; Hussaina, S.; Liu, G.W.; Shi, Z.Q.; Qiao, G.J. Magnetic Ni doping induced high power factor of Cu_2GeSe_3-based bulk materials. *J. Eur. Ceram. Soc.* **2021**, *41*, 3473–3479.
71. Lu, X.; Morelli, D.T.; Wang, Y.X.; Lai, W.; Xia, Y.; Ozolins, V. Phase stability, crystal structure, and thermoelectric properties of $Cu_{12}Sb_4S_{13-x}Se_x$ solid solutions. *Chem. Mater.* **2016**, *28*, 1781–1786.
72. Park, S.J.; Kim, I.H. Enhanced thermoelectric performance of Sn and Se double-doped famatinites. *J. Korean Phys. Soc.* **2023**, *83*, 57–64.

Disclaimer/Publisher's Note: The statements, opinions and data contained in all publications are solely those of the individual author(s) and contributor(s) and not of MDPI and/or the editor(s). MDPI and/or the editor(s) disclaim responsibility for any injury to people or property resulting from any ideas, methods, instructions or products referred to in the content.

Article

Flexible Active Peltier Coolers Based on Interconnected Magnetic Nanowire Networks

Tristan da Câmara Santa Clara Gomes, Nicolas Marchal, Flavio Abreu Araujo and Luc Piraux *

Institute of Condensed Matter and Nanosciences, Université Catholique de Louvain, Place Croix du Sud 1, 1348 Louvain-la-Neuve, Belgium
* Correspondence: luc.piraux@uclouvain.be

Abstract: Thermoelectric energy conversion based on flexible materials has great potential for applications in the fields of low-power heat harvesting and solid-state cooling. Here, we show that three-dimensional networks of interconnected ferromagnetic metal nanowires embedded in a polymer film are effective flexible materials as active Peltier coolers. Thermocouples based on Co-Fe nanowires exhibit much higher power factors and thermal conductivities near room temperature than other existing flexible thermoelectric systems, with a power factor for Co-Fe nanowire-based thermocouples of about 4.7 mW/K^{2}m at room temperature. The effective thermal conductance of our device can be strongly and rapidly increased by active Peltier-induced heat flow, especially for small temperature differences. Our investigation represents a significant advance in the fabrication of lightweight flexible thermoelectric devices, and it offers great potential for the dynamic thermal management of hot spots on complex surfaces.

Keywords: flexible thermoelectrics; active cooling; 3D nanowire networks

Citation: da Câmara Santa Clara Gomes, T.; Marchal, N.; Abreu Araujo, F.; Piraux, L. Flexible Active Peltier Coolers Based on Interconnected Magnetic Nanowire Networks. *Nanomaterials* **2023**, *13*, 1735. https://doi.org/10.3390/nano13111735

Academic Editors: Ting Zhang and Peng Jiang

Received: 21 March 2023
Revised: 19 May 2023
Accepted: 22 May 2023
Published: 25 May 2023

Copyright: © 2023 by the authors. Licensee MDPI, Basel, Switzerland. This article is an open access article distributed under the terms and conditions of the Creative Commons Attribution (CC BY) license (https://creativecommons.org/licenses/by/4.0/).

1. Introduction

Flexible thermoelectric (TE) materials and devices that easily harness the thermal energy of hot surfaces with complex geometries or even the human body for the conversion of heat into electricity offer innovative perspectives in a sustainable development context [1,2]. In particular, flexible TE generators should make it possible to respond to the rapid development of miniature, lightweight and functional portable electronic devices for nomadic products and medical sensors [3,4]. They add their own advantages, linked to their flexibility, lightness and conformability to the main attractions of TE converters, which are the absence of moving mechanical parts and operating noise, high reliability and an almost unlimited lifetime.

Flexible thermoelectrics are generally formed either from fully organic materials or from inorganic/organic hybrid systems, including composites with inorganic nanostructure fillers in a conducting polymer matrix and thin-film inorganic materials deposited on flexible polymer substrates [5–9]. Each of these TE systems requires the development of specific synthesis methods [10–21]. These include solution processes for the fabrication of conducting polymers, and physical vapor deposition, spin coating, screen printing, physical mixing and solution mixing for the fabrication of inorganic/organic hybrid systems. Although the thermoelectric properties of conducting polymers and their corresponding nanocomposites have been improved in recent decades, they are still significantly lower than those of bulk thermoelectric materials [5,6,8,9]. In addition, electrical and thermal contact resistances, as well as the mismatch in the thermal expansion coefficients of contacting materials, also affect the performance of such flexible TE devices [22].

Although flexible and lightweight TE devices are expected to generate the same interest in low-power miniaturized refrigeration based on the Peltier effect, it is only very recently that this field of application has received attention [23,24]. Flexible TE devices could enable the development of wearable devices for personalized thermoregulation with the prospect of drastically reducing the volume of heating or cooling by targeting the areas of the human body that require precise thermal doses [23–26]. In normal operation, Peltier coolers use materials with a high thermoelectric power factor and low thermal conductivity to reduce the reverse heat flow from the hot side. However, the increased development of electronic products, such as computer processors, LEDs and electric batteries, requires that conventional cooling systems be combined with dynamic thermal management to quickly dissipate heat peaks that interfere with the operation of the devices [27,28]. While heat removal from electronic components is usually achieved through high-thermal-conductivity materials that provide passive cooling, thermoelectric cooling has also been considered for dynamic thermal management [29]. Dynamic thermal management is a technique adopted at runtime to minimize hot spots and temperature peaks. Dynamic thermal management is of great interest because of its rapid adaptability to changing environments over time and the energy savings that it offers compared to passive materials. In this particular regime, called active cooling [30–32], the Peltier heat flow carried by the electric current is added to the natural flow of heat from the hot to the cold side. As recently demonstrated, materials suitable for the thermal management of electronic hot spots must not only have a high power factor but also high thermal conductivity [33]. Therefore, conventional Peltier TE modules are not suitable for this application due to the low thermal conductivity of their semiconductor components. On the contrary, some transition metals are excellent candidates as thermoelectric materials for active cooling. This is the case for cobalt, which, together with YbAl$_3$ [34], presents the highest room-temperature (RT) thermoelectric power factor among known materials, with a value of around 15 mW/K^{2}m, up to a factor of 10 higher than that of bismuth tellurium [35]. In addition, its thermal conductivity of ~100 W/Km is very high compared to that of Bi$_2$Te$_3$ (less than 2 W/Km). A new factor of merit, called the effective thermal conductivity, makes it possible to account for the flow of heat transported by materials during active cooling [30,33], i.e., accounting for both passive Fourier heat transport and active Peltier-driven heat flow. The effective thermal conductivity appears as the sum of the "passive" normal thermal conductivity and an "active" thermal conductivity that only starts up when the cooling power supplied by the Peltier module is activated. Using an active cooler made from a rigid bulk-type Co-CePd$_3$ thermoelectric couple, it was found that the effective thermal conductivity of the cooler can exceed 1000 W/Km in its active mode vs. 40 W/Km in the passive mode depending on the hot–cold temperature difference [30]. Although these recently obtained results open up new perspectives in the selection of thermoelectric materials useful for active cooling, hot spots appearing on arbitrarily shaped surfaces require the use of miniaturized, flat and flexible active Peltier coolers. In this context, the easy realization of high-performance flexible thermoelectric modules, with the advantages of reduced weight and size, remains challenging.

In this study, we demonstrate the ability to efficiently cool hot spots on electronic devices using lightweight, flexible Peltier coolers. We achieve this by developing an attractive electrodeposition method of manufacturing thermoelectric modules from ferromagnetic nanowire (NW) networks. Electrochemical synthesis has been shown to be a powerful method for fabricating multicomponent NWs with different metals due to its engineering simplicity, versatility and low cost [36–38]. In addition, NW networks have already been proposed as a pathway for efficient thermoelectric devices [39,40]. Herein, NW networks are obtained via simple electroplating within 3D porous polymer membranes so that the nanoconstituents active for Peltier cooling are completely embedded and perfectly interconnected within a polymer matrix (see the Experimental Section) [41–45]. In such centimeter-scale NW networks, electrical connectivity is essential to allow charge flow throughout the sample and to preserve the excellent bulk properties. The shapeable nanocomposite films also meet the key requirements for electrical, thermal and mechanical stability.

2. Results and Discussion

Flexible TE modules composed of p-type and n-type legs are fabricated via the successive electrodeposition of arrays of crossed nanowires (CNWs) in a single 22 µm thick polycarbonate (PC) template from a sputtered Au cathode by adapting a previously developed method [41,42,45] (see Figure 1a and the Experimental Section). Here, the TE film devices are made of electrodeposited n-type Co CNW and p-type Fe CNW legs. Next, the Au cathode is removed locally via plasma etching to create the multi-electrode architecture, as shown in Figure 1a,b. In this planar device, the TE elements are connected thermally in parallel and electrically in series, and the current is injected in the macroscopic direction of the film plane taking advantage of the very high degree of electrical connectivity of CNWs, as illustrated in Figure 1b. The device benefits from the mechanical properties of the porous polymer material, which allows for good flexibility and easy handling of the TE films, as shown in Refs. [45,46] (See Supporting Information, Video S1). Figure 1c shows a picture of an example of a nanowire-based active cooler, which highlights that a rigid support is not required to hold the self-supported thermoelectric film device. It should be noted that the dimensions of the device shown in Figure 1c are chosen to better present the characteristics of the system and do not correspond to those of the devices considered in this work, which have much shorter and wider legs to decrease the electrical resistance and increase the performance of the nanowire-based active cooler (see Section 4). The widths of the p and n legs can be selected individually during the successive electrodeposition processes, allowing for thermal and electrical impedance matching, while the lengths of both nanowire legs can be adjusted during the etching process. The fabrication method using track-etch technology allows for thicker nanocomposite films (up to 100 µm) while maintaining flexibility. Note that this system offers much better mechanical properties (in terms of softness and flexibility) than metallic films of similar thicknesses. The polymer matrix also protects the metal NWs from oxidation. In addition, the non-toxicity of the constituent materials makes them wearable devices, whereas toxicity is an issue for many existing flexible thermoelectric devices, apart from conductive polymers [2,3]. The scanning electron microscopy images (see Figure 1d,e) obtained after the complete dissolution of the PC membrane of free-standing CNWs with a 105 nm diameter and a 20% packing density reveal the interconnections between the NWs.

The TE power factors (PF = S^2/ρ, where ρ is the electrical resistivity, and S is the Seebeck coefficient) of the n-type Co CNW and p-type Fe CNW legs are determined separately using previously developed experimental setups for measuring electrical resistance and thermopower as a function of temperature (See Methods and Supporting Information Section S1, Figures S1 and S2). The RT resistivities of the Co and Fe CNW networks are estimated to be 7.1 µΩcm and 12.8 µΩcm, respectively. These values are significantly larger than those of the bulk metals (5.8 µΩcm and 9.8 µΩcm for Co and Fe, respectively). At RT, the resistivity of a bulk metal is fully dominated by large-angle electron–phonon scattering [47]. The residual resistivity due to defect scattering is typically more than a hundred times lower than the resistivity obtained at $T = 300$ K. On the contrary, in metal NWs, the residual resistivity increases strongly, as evidenced by the reduced residual resistivity ratio (RRR) observed in Co and Fe NWs being close to 4 and 5, respectively (see Supporting Information Section S1). Such an increase in residual resistivity results from an increase in electron scattering caused by structural defects in polycrystalline NWs made using electrodeposition [48] and the surface of the NWs. In particular, it was found for polycrystalline gold NWs that the increase in resistivity due to the external surface scattering of the electrons is not dominant until the diameter is very close to the grain size (40–50 nm) [49]. Therefore, contrary to ultra-small-diameter NWs and thin metal films [50,51], size effects are not very pronounced in our NW system due to the relatively large diameter of the NW (105 nm), so scattering at the grain boundaries is mainly responsible for the increased resistivity of the metal NWs compared to the bulk materials. Furthermore, the temperature dependence and measured Seebeck coefficient values of the NW networks correspond to those of bulk ferromagnets (see See Methods and Supporting Information Section S1,

Figure S2). The measured RT values of S for the Co and Fe CNW networks are −28 µV/K and +15 µV/K, respectively, values very close to those of the bulk metals (−30 µV/K and +15 µV/K for Co and Fe, respectively).

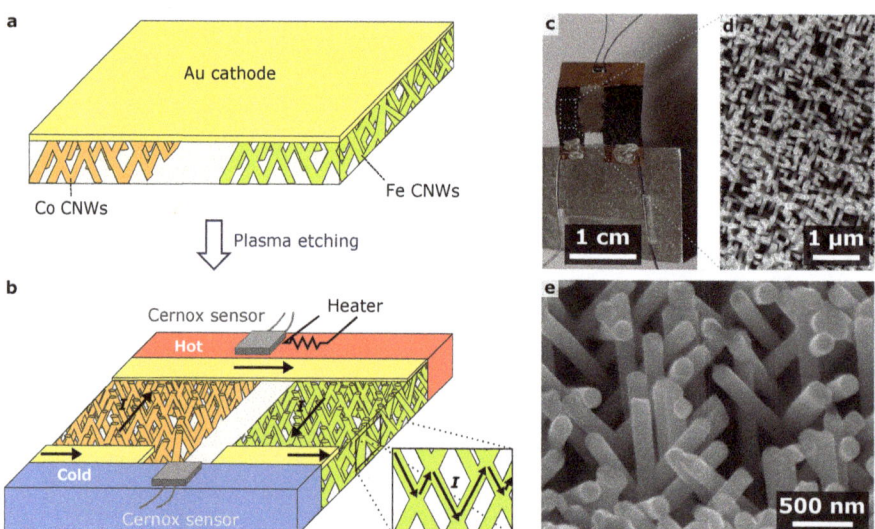

Figure 1. Flexible thermoelectric coolers based on interconnected nanowire networks. (**a**,**b**) Fabrication technique of a thermoelectric device consisting of p- and n-type interconnected metallic nanowire networks (with Fe for p-type and Co for n-type materials). The Co and Fe NW networks, shown in orange and green respectively, are obtained via direct electrodeposition from a Au cathode within a three-dimensional porous polycarbonate membrane (**a**). The thermocouple is made after local removal of the Au electrode via plasma etching (**b**). The temperature at the thermoelectric junction can be raised by means of a resistive heater. Cernox resistance thermometers are used to determine the temperatures of the cold and hot junctions. The inset in (**b**) shows the current path in the nanowire networks. (**c**) Picture of the nanowire-based active cooler corresponding to the schematics in (**b**), showing the flexible and self-supported device. (**d**,**e**) Scanning electron microscopy images of self-supported interconnected Co nanowires with 105 nm diameter showing a 50°-tilted view of the macroscopic nanowire network film (**d**) and the nanowire branched structure at higher magnification (**e**).

Consequently, extremely high PF values of ∼11.0 mW/K²m and ∼1.8 mW/K²m at RT are estimated for the Co and Fe CNWs, respectively [45]. These values are only slightly lower than the bulk values, PF ∼15.0 mW/K²m and ∼2.3 mW/K²m for Co and Fe, respectively [35,52,53]. The PF values obtained for these magnetic NWs are even larger than those of the widely used TE material bismuth telluride (in the range of 1–6 mW/K²m [54]) and at least one order of magnitude larger than those reported for flexible thermoelectric films based on optimized conducting polymers and inorganic/organic hybrid systems [6,9,12]. Using these values, the power factor of the Co-Fe nanowire-based thermocouple can be estimated as follows [32,55]:

$$\text{PF}_{\text{Co-Fe}} = \frac{(S_{\text{Fe}} - S_{\text{Co}})^2}{(\sqrt{\rho_{\text{Fe}}} + \sqrt{\rho_{\text{Co}}})^2}. \tag{1}$$

At room temperature, the power factor is about 4.7 mW/K²m.

The complete characterization of the properties of nanostructured thermoelectric materials, including the measurement of thermal conductivity, is generally complex and requires the development of specific integrated measurement devices [56–58]. As pointed out above, materials suitable for active cooling must have high thermal conductivity κ, in contrast to traditional thermoelectric materials, which require low thermal conductivity to achieve a high figure of merit $Z = PF/\kappa$. Because of the very low thermal conductivity of PC ($\kappa = 0.2$ W/Km at RT) and the large packing factor of the NW networks, the contribution of the polymer matrix to heat transport is much smaller than that of the metallic NWs. Although the measurement of the thermal conductivity of isolated NWs and NW networks is very delicate, it has been the subject of several works (for recent reviews, see Refs. [56,57,59]). For metallic NWs, heat is predominantly transported by electrons, and the Wiedemann–Franz law ($\kappa\rho = L_0 T$, where $L_0 = 2.44 \ 10^{-8}$ V^2/K^2 is the Lorenz number) holds at room temperature to a good approximation, as demonstrated in previous studies on Ni and Ag NWs [60–62]. Thus, the thermal conductivity of the Co and Fe NWs can be estimated at RT from their resistivity values, leading to $\kappa \sim 100$ W/Km and $\kappa \sim 56$ W/Km for the Co CNWs and Fe CNWs, respectively. Although these thermal conductivity values are somewhat lower than those of the bulk metals, they are up to two orders of magnitude higher than those of the flexible thermoelectrics developed so far [3,5,63]. It should be noted that the high thermal conductivity of the NW-based nanocomposite system in any direction in the plane of the film, as well as in the direction perpendicular to the film, is directly related to the perfect electrical and thermal connectivity between the NWs along the entire length of the film. In conventional nanocomposite systems consisting of randomly distributed NWs in a polymer matrix, the high thermal contact resistances between the constituents combined with the low thermal conductivity of the polymer lead to significantly lower thermal conductivities. A further comparison of the power factor and thermal conductivity of ferromagnetic CNW networks with those of several flexible and rigid thermoelectric materials is shown in Figure 2a. From these results, it appears that ferromagnetic CNWs are suitable as active TE coolers, as this particular application requires materials with both high PF and high thermal conductivity [30,33]. Based on the thermal conductivity estimates made above, the ZT figure of merit of Co and Fe nanowire networks at room temperature is about 3×10^{-2} and 10^{-2}, respectively. The drawings in Figure 2b,c compare active cooling and refrigeration applications. In the case of active cooling (see Figure 2b), both Peltier flow and Fourier heat conduction are directed from the hot side to the cold side, in contrast to the case of refrigeration (see Figure 2c), where the two heat flows oppose each other.

Two CNW thermocouples with different network heights and resistive characteristics are reported here. In sample 1, the PC template is partially filled by the NWs (Figure 3a), thus showing a higher resistance ($R = 32.4$ mΩ) than sample 2, where the metallic materials completely fill the pores (Figure 3b, $R = 23.6$ mΩ). The thermocouple legs are approximately 3 mm long, and the widths are set to 22 mm and 40 mm for the Co and Fe CNW legs, respectively, to achieve reasonably good thermal and electrical impedance matching. The two extremities of the thermocouple are connected to the heat sink maintained at RT, while the heated thermocouple junction is physically attached to a Cernox temperature sensor placed in close contact with a resistive heating element, as shown in the insets in Figure 3a,b. The Peltier characteristics of the two samples are first extracted with the resistive heater turned off and by injecting different current values through the sample, which induce both Peltier cooling/heating and Joule heating. The steady-state temperature gradient recorded using the Cernox sensor $\Delta T = T - T_0$, where T and T_0 denote the temperatures measured at the thermocouple junction and heat sink, respectively, is reported in Figure 3a,b for the high- and low-resistance samples, respectively. As can be seen, the data (black circles in Figure 3a,b) correspond well to the expected variation due to the Peltier effect contribution ($\Delta T \propto I$) and Joule heating contribution ($\Delta T \propto I^2$), as shown by the dashed red curves in Figure 3a,b. These combined effects lead to an asymmetric variation in temperature depending on the current direction. The optimal current that minimize ΔT can be estimated from the theoretical expression $I_{opt} = ST/R$. Considering a sensitivity of 45 µV/K at RT for the Co-Fe thermocouple, this leads to $I_{opt} = -420$ mA (sample 1) and $I_{opt} = -590$ mA (sample 2). The experimental results obtained for both samples are in good agreement with these predictions, as shown in Figure 3a,b.

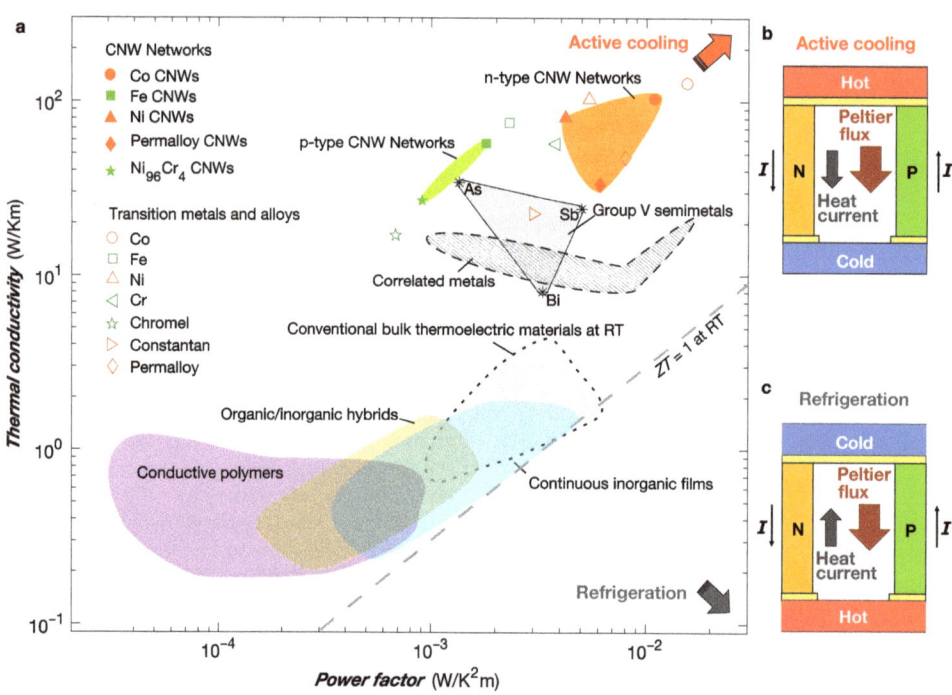

Figure 2. Comparison of the active cooling performance among several flexible thermoelectric materials. (**a**) Thermoelectric power factor PF vs. thermal conductivity κ for flexible TE systems made of n-type (red symbols) and p-type (green symbols) magnetic nanowire networks [45], conductive polymers, organic/inorganic hybrids and continuous inorganic films [1–3]. The data for conventional bulk thermoelectric materials near room temperature [64–69], correlated metals [34,70,71], group V semimetals [72] and transition metals and alloys [30,35,52,53,73–75] are also presented. The gray dashed line shows the case $ZT = 1$ (with $Z = PF/\kappa$ as the figure of merit), which corresponds to the best power conversion efficiency achievable to date. The most suitable active Peltier coolers are located at the top right of the graph, as indicated by the red arrow. In contrast, the most suitable materials for conventional Peltier refrigeration are located at the bottom right of the graph, as indicated by the gray arrow. (**b**,**c**) Schematic drawings showing the differences between active cooling (**b**) and refrigeration (**c**). In the active cooling mode, Peltier heat flows from the hot side to the cold side, increasing Fourier heat conduction rather than opposing it as in the refrigeration mode.

Figure 3c,d show the temperature versus time changes measured using the Cernox sensor for the two samples according to the direction of the DC current supplied to the NW-based Peltier device. After 100 s, a negative current corresponding to the optimal current is injected in the thermocouple for 400 s, leading to a net cooling at the junctions of $\Delta T_- = -0.27$ K and $\Delta T_- = -1.12$ K for samples 1 and 2, respectively. Then, the current is turned off for 400 s to restore the working temperature of 300 K. Next, changing the direction of the optimal current supplied to the Peltier device leads to temperature increases of $\Delta T_+ = +0.81$ K and $\Delta T_+ = +3.35$ K for samples 1 and 2, respectively. The Peltier and Joule contributions at the optimal current, ΔT_P and ΔT_J, can be extracted as given by $\Delta T_P = (\Delta T_+ - \Delta T_-)/2 = \pm 0.54$ K (± 2.23 K) and $\Delta T_J = (\Delta T_+ + \Delta T_-)/2 = 0.27$ K (1.12 K) for the high- (low-) resistance samples, respectively. As can be seen, this sequence applied to the NW-based thermocouple has a fast response and can be switched quickly from heating to cooling.

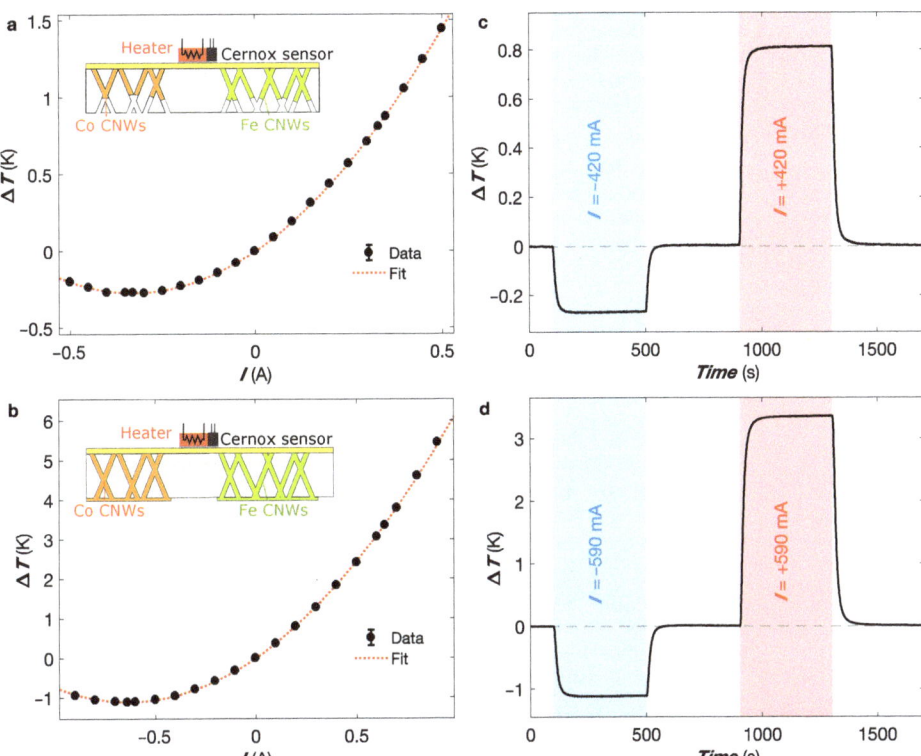

Figure 3. Device characterization as a function of current injection. (**a**,**b**) Measured total temperature changes at the Peltier junction of the NW-based thermocouple versus current intensity applied both forward and reverse for samples 1 (**a**) and 2 (**b**), respectively. The dotted lines provide a fit, including Joule and Peltier contributions to temperature variations. The insets in (**a**,**b**) show schematics of the hot side of thermoelectric coolers consisting of networks of interconnected Co and Fe nanowires of 105 nm in diameter partially filling a porous polymer film (**a**), with the same nanowire constituents completely filling the porous medium with the formation of a thin metallic layer on the surface in (**b**). (**c**,**d**) Temperature versus time traces of the sum of the Joule and Peltier heats relative to a working temperature of 300 K, as recorded using the Cernox sensor. The direct currents of 420 mA for sample 1 (**c**) and of 590 mA for sample 2 (**d**) are applied sequentially forward and reverse in the NW-based thermocouple. Error bars in (**a**,**b**) are smaller than the markers, reflecting the uncertainty of the voltage and temperature measurements, and they are set to two times the standard deviation, gathering 95% of the data variation (See Supporting Information Section S3 for details).

Following the measurement scheme shown in Figures 1a and 3a,b, the thermal conductance of the thermoelectric device with a heat sink at ambient temperature is evaluated by heating the thermocouple junction using the heat source provided by the resistive heater and measuring the temperature difference $\Delta T = T - T_0$ after establishing a steady-state temperature distribution. For the passive cooling ($I = 0$) regime, the proportional relationship between the known power input Q and the resulting temperature drop is shown in Figure 4a,b for both samples. The same measurements are performed in the active cooling regime using the optimum currents for sample 1 ($I_{opt} = -420$ mA) and sample 2 ($I_{opt} = -590$ mA). As expected, the active Peltier effect leads to smaller ΔT values for a given power dissipation in the heater than the passive cooling state. In contrast to the passive mode, the relationship between Q and ΔT in the active mode is no longer linear in particular for small temperature differences, indicating an increase in thermal conductance, as given by $Q/\Delta T$ (see insets in Figure 4a,b). Figure 4c shows the temperature versus time traces obtained during the successive switching on of the thermal load on the thermoelectric junction and the optimal cooling currents of samples 1 (in black) and 2 (in yellow). For sample 1, the temperature increase resulting from a power dissipation $P = 0.5$ mW in the resistive heater can be completely compensated by the NW-based TE cooler operating at its optimum current of -420 mA (Figure 4c in black). Similarly, for sample 2, a ΔT of 1.1 K resulting from a thermal load of 1.2 mW in the heater can be canceled out by injecting an optimal Peltier current of -590 mA (Figure 4c in yellow).

At the optimal current I_{opt}, the ratio between the effective thermal conductance K_{eff} (including both the passive cooling and active Peltier contributions—see Supporting Information Section S2) and the thermal conductance of the thermocouple is given by [30,33]

$$\frac{K_{eff}}{K} = \left[\frac{1}{2}\frac{(S_p - S_n)^2 T_H^2}{KR\Delta T} + 1\right], \qquad (2)$$

where S_p and S_n are the respective Seebeck coefficients for the p and n legs, R is the electrical resistance, and T_H is the hot-side temperature. This expression indicates that K_{eff}/K increases when the temperature difference is low, in agreement with the results reported in Figure 4d for samples 1 and 2. As an example, for $\Delta T = 0.1$ K, the gain on the thermal conductance due to active cooling reaches factors 4 and 12 for samples 1 and 2, respectively. It should be noted that a decrease in contact resistance, which can hardly be made negligible compared to that of the conductive film formed by the NW array (see Section 4), should lead to even better performance. Another limiting factor is radiation losses, which are not evaluated in this study.

For metallic legs, Equation (2) can be simplified as follows:

$$\frac{K_{eff}}{K} = \left[\frac{1}{2}\frac{(S_p - S_n)^2 T_H}{L_0 \Delta T} + 1\right], \qquad (3)$$

where L_0 is the Lorenz number, equal to the ratio KR/T in the Wiedemann–Franz law. Equation (3) makes it easy to estimate, for a given metallic thermocouple, the maximum thermal conductance gain at any working temperature for various values of ΔT. It should be noted that Equation (3) does not take into account additional heat losses due to radiation or the contribution of contact resistances that reduce the K_{eff}/K ratio, as observed in our Co-Fe CNW thermocouples (see Figure 4d). For a Co-Fe thermocouple, this leads to $K_{eff}/K \approx 13$ for $\Delta T = 1$ K at RT. Therefore, the effective thermal conductivities of the Co and Fe legs with bulk values of $\sim$125 W/Km and $\sim$80 W/Km, respectively, both exceed 1000 W/Km for $\Delta T = 1$ K, which corresponds to the values of the best thermal conductors at RT. In addition, a larger improvement can be obtained when the temperature of the hot spot is higher than RT. From Equation (3), it is also seen that active Peltier cooling removes more heat than passive cooling if $\Delta T < (S_p - S_n)^2 T_H/2L_0$, which corresponds to $\Delta T < 12.4$ K for a Co-Fe thermocouple at ambient temperature. Furthermore, as the thermal conductivity of Cu is about four times higher than that of Co NWs, it can be seen from Equation (3) that, under ideal conditions and for a ΔT lower than about 3.8 K, the Co-Fe NW thermocouple has a better cooling capacity in active mode than passive heat transfer via Cu. Moreover, it should be noted that the main advantages of active Peltier cooling are the dynamic and controllable aspects, which allow for dynamic heat removal for precise temperature control. In a regime of rapid temperature increases in small amounts (i.e., small ΔT), the active Peltier cooler can effectively increase the effective thermal conductance of the cooling system, which can be precisely controlled by adjusting the electric current flowing through the thermocouple. Therefore, the large improvement in effective thermal conductance in our flexible Co-Fe NW-based devices holds promise for the accurate and fast-response temperature management of hot spots in electronic devices. Furthermore, by simply reversing the polarity of the current, the effective conductance can be significantly reduced, allowing for the easy realization of flexible thermal switches, which have great potential for heat management; thermal logic devices; and magnetocaloric/electrocaloric coolers [76,77].

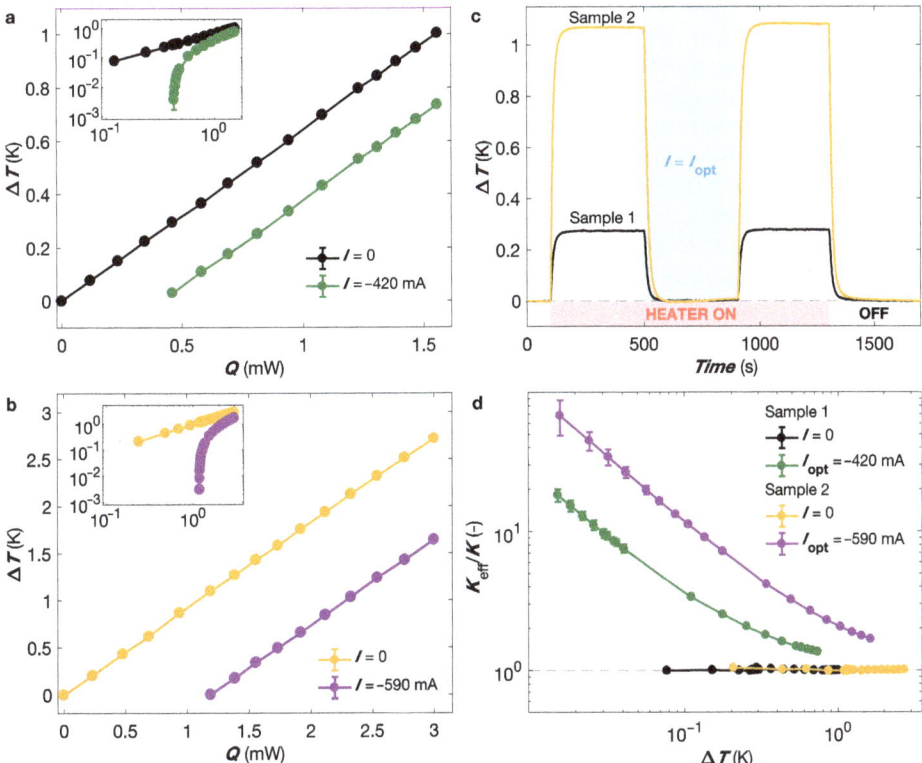

Figure 4. Active cooling of an electronic device. (**a**,**b**) Temperature variations of the Cernox sensor as a function of thermal load when optimal electric currents and zero current are passed through the partially filled (**a**) and completely filled (**b**) Co and Fe nanowire thermocouples. The slopes of the lines provide the thermal conductance. Insets: ΔT vs. Q plots on log-log scales. (**c**) The time dependence of temperature changes during successive switching on of the thermal load on the thermoelectric junction and the optimal cooling current of samples 1 (in black, $Q = 0.5$ mW and $I_{opt} = -420$ mA) and 2 (in yellow, $Q = 1.2$ mW and $I_{opt} = -590$ mA). (**d**) Variation in effective thermal conductance K_{eff} normalized by the passive-mode thermal conductance K with the temperature gradient ΔT for partially filled and completely filled Co and Fe nanowire thermocouples at the optimal switching currents. Error bars in (**a**,**b**,**d**) reflect the uncertainty of the voltage and temperature measurements, and they are set to two times the standard deviation, gathering 95% of the data variation (See Supporting Information Section S3 for details).

3. Conclusions

In this paper, we propose a simple, versatile and reliable fabrication technique for lightweight, non-toxic, flexible and shapeable TE devices consisting of interconnected nanowire arrays embedded in 22 µm thick polymer films. The combination of a high power factor and high thermal conductivity in ferromagnetic nanowires provides flexible active Peltier coolers with significantly better performance than other existing flexible

TE systems. A power factor of about 4.7 mW/K^{2}m was obtained at room temperature for Co-Fe nanowire-based thermocouples, providing the highest value reported to date for flexible TE films. We found that the ratio of the effective thermal conductance of Co-Fe nanowire-based thermocouples in the active mode to the thermal conductance in the passive mode increases sharply as the hot–cold temperature difference decreases. The process of integrating electrochemically produced nanowire thermocouples into active TE coolers is simple to implement. The size of the planar device and thermocouples, as well as the number of modules, can be adjusted and optimized to cool hot spots on surfaces with complex geometries. Another interesting prospect of 3D nanowire networks is the possibility to fabricate multilayer nanowires to achieve magnetic control of the thermoelectric properties [41,46]. It should be noted that a route for the fabrication of vertical thermocouples based on CNWs has very recently been proposed [46]. This architecture, which is closer to that generally found in usual thermoelectric devices and offers excellent prospects for flexible active coolers, requires further optimization to limit contact resistance issues. These results represent a significant advance in the design of flexible TE device assemblies and offer real prospects for the thermal management of electronic hot spots.

4. Experimental Section

4.1. Fabrication of Flexible Nanowire-Based TE Devices

Polycarbonate (PC) porous membranes with interconnected pores were fabricated by exposing a 22 μm thick PC film to a two-step irradiation process [78,79]. The topology of the membranes was defined by exposing the film to an initial irradiation step at two fixed angles of −25° and +25° with respect to the normal axis of the film plane. After rotating the PC film in the plane by 90°, the second irradiation step took place at the same fixed angular irradiation flux to finally form a 3D nanochannel network. Then, latent tracks were chemically etched following a previously reported protocol [80] to obtain 3D porous membranes with pores 105 nm in diameter and a volumetric porosity of 20%. Next, the PC templates were coated on one side using an e-beam evaporator with a metallic Cr (3 nm)/Au (700 nm) bilayer to serve as the cathode during electrochemical deposition.

Flexible NW-based TE devices were manufactured via the successive electroplating of n-type Co NW legs and p-type Fe NW legs. Electrodeposition was performed at RT in the potentiostatic mode using a Ag/AgCl reference electrode and a Pt counter electrode. The electrodeposition of Fe and Co CNWs into interconnected pore PC templates was carried out at the respective constant potentials of −1.20 V and −0.95 V using the following electrolytes: 0.5 M $FeSO_4 \cdot 7 H_2O$ + 0.5 M H_3BO_3 at pH 2, and 0.5 M $CoSO_4 \cdot 7 H_2O$ + 0.5 M H_3BO_3 at pH 3.6.

The dimensions of the p and n NW legs were 40 mm × 3 mm and 22 mm × 3 mm in area, respectively, and they were 22 μm thick when pore filling was complete. After the electroplating process, a single etching step carried out using plasma etching through a mechanical mask allowed for the local removal of the surface gold layer to finalize the design of the planar TE module. At the end of this etching step, only metallic segments of the continuous gold layer remained, which allowed for the thermoelectric legs p and n to be alternately connected in series.

4.2. Electrical and Thermoelectric Measurements

Thermoelectric and resistivity measurements of individual Co NW and Fe NW networks were performed as a function of temperature using homemade setups. For conducting electrical and thermoelectric transport measurements, the cathode was locally removed via plasma etching to create a two-probe design suitable for electric measurements (see Supporting Information Section S1, Figure S1). In this configuration, the current was directly injected to the branched CNW structure of the film samples from unetched sections of the metallic cathode, where the electrical contacts were directly made using Ag paint, and it passed through the NW network thanks to the high degree of electrical connectivity of the CNWs. The typical resistance values of the prepared specimens were in the range of a few tens of ohms. For each sample, the input power was kept below 0.1 µW to avoid self-heating, and the resistance was measured within its ohmic resistance range with a resolution of one part in 10^5. The RT resistivities of the Co and Fe NW networks were estimated from low-temperature resistance measurements assuming that the Matthiesen's rule holds. In this case, the resistivity at RT is given by $\rho_{NWs}^{RT} = \rho_{FM}^{RT} + \rho_{NWs}^{0}$, where ρ_{FM}^{RT} is the resistivity of the FM that composes the NWs at RT due to thermally excited scatterings, and ρ_{NWs}^{0} is the residual resistivity of the NWs due to impurities, and surface and grain-boundary scatterings. For an NW diameter that is not too small ($\phi \leq 40$ nm), the thermally induced scattering effects are independent from the sample dimensions, nanostructuration and defect concentration [81]. Therefore, ρ_{FM}^{RT} can be taken as the ideal resistivity value at RT reported for bulk materials; i.e., $\rho_{Co}^{RT} = 5.8$ µΩcm and $\rho_{Fe}^{RT} = 9.8$ µΩcm. Moreover, the resistivity of the NW networks at $T = 10$ K can be approximated to $\rho_{NWs}^{10\,K} \sim \rho_{NWs}^{0}$. Finally, using the measured residual resistivity ratio RRR = $R_{NWs}^{RT}/R_{NWs}^{10\,K} \sim (\rho_{FM}^{RT} + \rho_{NWs}^{0})/\rho_{NWs}^{0}$ (see Supporting Information Section S1, Figure S2a), the RT resistivity of the NWs can be estimated as $\rho_{NWs}^{RT} \sim \rho_{FM}^{RT}$RRR/(RRR − 1). The resistivity values obtained for the CNW systems ($\rho_{Co}^{RT} = 7.1$ µΩcm and $\rho_{Fe}^{RT} = 12.8$ µΩcm) were slightly larger than those obtained for the bulk materials, as expected for electrodeposited nanostructured materials.

The thermoelectric power was measured by attaching one end of the sample to a copper sample holder using silver paint and a resistive heater to the other end (see Supporting Information Section S1, Figure S1). The voltage leads were made of thin Chromel P wires, and the contribution of the leads to the measured thermoelectric power was subtracted using the recommended values for the absolute thermopower of Chromel P. The temperature gradient was monitored with a small-diameter type E differential thermocouple. A typical temperature difference of 1 K was used in the measurements. Electrical and thermoelectric measurements were performed under vacuum. The temperature of the samples can be varied from 10 to 320 K.

Active Peltier cooling experiments were conducted at room temperature on two different NW-based thermocouples made of interconnected Co and Fe NWs 105 nm in diameter. In the first NW-based thermocouple, the NW network partially filled the 22 µm thick porous polymer film, with a resistance of 32.4 mΩ. In the second specimen, the same nanowire constituents completely filled the porous medium with the formation of a thin metallic layer on the surface, thus leading to a lower resistance of 23.6 mΩ. Through independent characterization tests, the electrical contact resistance was estimated at ∼6 ± 2 mΩ, thus being significantly smaller, although not negligible, than the resistance of the two samples investigated in this study. The relatively low contact resistance can be attributed to the large surface area of the electrode used as a cathode for direct nanowire growth via electrodeposition combined with the interconnected structure of the nanowire network and its high packing factor (∼20%). The heat absorbed or released at the thermoelectric junction, whose temperature can be raised by means of a resistive heater, was measured using a small Cernox thin-film resistance sensor (<3 mg, 1 mm^2; Cernox 1010, Lake Shore Cryotronics Inc., USA) attached at the junctions between the NW networks. The temperature resolution of this highly sensitive thermometer is about 1 mK, which enables the detection of Peltier-effect-based heating or cooling, while a DC electrical current was applied sequentially forward and reverse in the NW-based thermocouples.

Supplementary Materials: The following supporting information can be downloaded at: https://www.mdpi.com/article/10.3390/nano13111735/s1, Figure S1: Device configuration for measurements of the resistance and the Seebeck coefficient of the metallic NW networks. a, Schematic of 3D interconnected nanowire network film grown by electrodeposition from a Au cathode into a 22 µm thick polycarbonate template with crossed-nanopores. b, Two-probe electrodes design obtained by local etching of the Au cathode. c, The voltage differential ∆V induced by the injected current I between the two metallic electrodes is measured while the two electrodes are maintained at an identical and constant temperature. d, Heat flow is generated by a resistive element at one electrode while the other electrode is maintained at desired temperature. The temperature difference ∆T between the two metallic electrodes is measured by a thermocouple while thermoelectric voltage ∆V settles. Figure S2: Temperature variation of the electrical resistance and Seebeck coefficient of the Co and Fe CNW networks. a, Measured resistance vs. temperature curves for the interconnected Co and Fe NW networks, 105 nm in diameter, 10 mm × 2.5 mm × 0.022 mm (thickness) in sizes. b, Measured $S(T)$ curves obtained on the same samples. The data are compared with the bulk values indicated by the dashed lines. Error bars are smaller than the markers, reflecting the uncertainty of the voltage and temperature measurements and set to two times the standard deviation, gathering 95% of the data variation. Video S1: Movie showing the flexible thermoelectric active cooler composed of three p-n modules. The electrical resistance of the nanowire nanocomposite remains unchanged when subjected to twisting movements. References [82,83] are cited in the supplementary materials.

Author Contributions: T.d.C.S.C.G. performed most of the experiments, analyzed the data and contributed to the writing of the manuscript. N.M. performed the experiments, analyzed the data and contributed to the writing of the manuscript. F.A.A. analyzed the data. L.P. contributed to the initial ideas, analyzed the data and contributed to the writing of the manuscript. All authors have read and agreed to the published version of the manuscript.

Funding: Financial support was provided by the Wallonia/Brussels Community (ARC 18/23-093) and the Belgian Fund for Scientific Research (FNRS).

Data Availability Statement: The data presented in this study are available on request from the corresponding author.

Acknowledgments: N.M. acknowledges the Research Science Foundation of Belgium (FRS-FNRS) for financial support (FRIA grant). F.A.A. is a Research Associate at the FNRS. The authors thank E. Ferain and the it4ip Company for supplying polycarbonate membranes. The authors would like to thank Pascal Van Velthem for his technical assistance in making the devices.

Conflicts of Interest: The authors declare no conflict of interest.

References

1. Fan, Z.; Zhang, Y.; Pan, L.; Ouyang, J.; Zhang, Q. Recent developments in flexible thermoelectrics: From materials to devices. *Renew. Sustain. Energy Rev.* **2021**, *137*, 110448. [CrossRef]
2. Zhang, L.; Shi, X.L.; Yang, Y.L.; Chen, Z.G. Flexible thermoelectric materials and devices: From materials to applications. *Mater. Today* **2021**, *46*, 62–108. [CrossRef]
3. Wang, Y.; Yang, L.; Shi, X.L.; Shi, X.; Chen, L.; Dargusch, M.S.; Zou, J.; Chen, Z.G. Flexible Thermoelectric Materials and Generators: Challenges and Innovations. *Adv. Mater.* **2019**, *31*, 1807916. [CrossRef] [PubMed]
4. Cao, T.; Shi, X.L.; Chen, Z.G. Advances in the design and assembly of flexible thermoelectric device. *Prog. Mater. Sci.* **2023**, *131*, 101003. [CrossRef]
5. Bahk, J.H.; Fang, H.; Yazawa, K.; Shakouri, A. Flexible thermoelectric materials and device optimization for wearable energy harvesting. *J. Mater. Chem. C* **2015**, *3*, 10362–10374. [CrossRef]
6. Du, Y.; Xu, J.; Paul, B.; Eklund, P. Flexible thermoelectric materials and devices. *Appl. Mater. Today* **2018**, *12*, 366–388. [CrossRef]
7. Lin, S.; Zhang, L.; Zeng, W.; Shi, D.; Liu, S.; Ding, X.; Yang, B.; Liu, J.; Lam, K.h.; Huang, B.; et al. Flexible thermoelectric generator with high Seebeck coefficients made from polymer composites and heat-sink fabrics. *Commun. Mater.* **2022**, *3*, 44. [CrossRef]
8. Masoumi, S.; O'Shaughnessy, S.; Pakdel, A. Organic-based flexible thermoelectric generators: From materials to devices. *Nano Energy* **2022**, *92*, 106774. [CrossRef]
9. Li, Y.; Lou, Q.; Yang, J.; Cai, K.; Liu, Y.; Lu, Y.; Qiu, Y.; Lu, Y.; Wang, Z.; Wu, M.; et al. Exceptionally High Power Factor Ag2Se/Se/Polypyrrole Composite Films for Flexible Thermoelectric Generators. *Adv. Funct. Mater.* **2022**, *32*, 2106902. [CrossRef]
10. Scholdt, M.; Do, H.; Lang, J.; Gall, A.; Colsmann, A.; Lemmer, U.; Koenig, J.D.; Winkler, M.; Boettner, H. Organic Semiconductors for Thermoelectric Applications. *J. Electron. Mater.* **2010**, *39*, 1589–1592. [CrossRef]
11. Kim, G.H.; Shao, L.; Zhang, K.; Pipe, K.P. Engineered doping of organic semiconductors for enhanced thermoelectric efficiency. *Nat. Mater.* **2013**, *12*, 719–723. [CrossRef] [PubMed]
12. Bubnova, O.; Khan, Z.U.; Malti, A.; Braun, S.; Fahlman, M.; Berggren, M.; Crispin, X. Optimization of the thermoelectric figure of merit in the conducting polymer poly(3,4-ethylenedioxythiophene). *Nat. Mater.* **2011**, *10*, 429–433. [CrossRef] [PubMed]
13. Bubnova, O.; Crispin, X. Towards polymer-based organic thermoelectric generators. *Energy Environ. Sci.* **2012**, *5*, 9345–9362. [CrossRef]
14. Yakuphanoglu, F.; Şenkal, B.F.; Saraç, A. Electrical Conductivity, Thermoelectric Power, and Optical Properties of Organo-Soluble Polyaniline Organic Semiconductor. *J. Electron. Mater.* **2008**, *37*, 930–934. [CrossRef]
15. Tang, X.; Liu, T.; Li, H.; Yang, D.; Chen, L.; Tang, X. Notably enhanced thermoelectric properties of lamellar polypyrrole by doping with β-naphthalene sulfonic acid. *RSC Adv.* **2017**, *7*, 20192–20200. [CrossRef]
16. Du, Y.; Shen, S.Z.; Cai, K.; Casey, P.S. Research progress on polymer–inorganic thermoelectric nanocomposite materials. *Prog. Polym. Sci.* **2012**, *37*, 820–841. [CrossRef]

17. Yao, Q.; Chen, L.; Zhang, W.; Liufu, S.; Chen, X. Enhanced Thermoelectric Performance of Single-Walled Carbon Nanotubes/Polyaniline Hybrid Nanocomposites. *ACS Nano* **2010**, *4*, 2445–2451. [CrossRef]
18. Meng, C.; Liu, C.; Fan, S. A Promising Approach to Enhanced Thermoelectric Properties Using Carbon Nanotube Networks. *Adv. Mater.* **2010**, *22*, 535–539. [CrossRef]
19. Du, Y.; Cai, K.F.; Chen, S.; Cizek, P.; Lin, T. Facile Preparation and Thermoelectric Properties of Bi2Te3 Based Alloy Nanosheet/PEDOT:PSS Composite Films. *ACS Appl. Mater. Interfaces* **2014**, *6*, 5735–5743. [CrossRef]
20. Wan, C.; Gu, X.; Dang, F.; Itoh, T.; Wang, Y.; Sasaki, H.; Kondo, M.; Koga, K.; Yabuki, K.; Snyder, G.J.; et al. Flexible n-type thermoelectric materials by organic intercalation of layered transition metal dichalcogenide TiS2. *Nat. Mater.* **2015**, *14*, 622–627. [CrossRef]
21. Tian, R.; Wan, C.; Wang, Y.; Wei, Q.; Ishida, T.; Yamamoto, A.; Tsuruta, A.; Shin, W.; Li, S.; Koumoto, K. A solution-processed TiS_2/organic hybrid superlattice film towards flexible thermoelectric devices. *J. Mater. Chem.* **2017**, *5*, 564–570. [CrossRef]
22. He, R.; Schierning, G.; Nielsch, K. Thermoelectric Devices: A Review of Devices, Architectures, and Contact Optimization. *Adv. Mater. Technol.* **2018**, *3*, 1700256. [CrossRef]
23. Hong, S.; Gu, Y.; Seo, J.K.; Wang, J.; Liu, P.; Meng, Y.S.; Xu, S.; Chen, R. Wearable thermoelectrics for personalized thermoregulation. *Sci. Adv.* **2019**, *5*, eaaw0536. [CrossRef] [PubMed]
24. Xu, S.; Li, M.; Dai, Y.; Hong, M.; Sun, Q.; Lyu, W.; Liu, T.; Wang, Y.; Zou, J.; Chen, Z.G.; et al. Realizing a 10 °C Cooling Effect in a Flexible Thermoelectric Cooler Using a Vortex Generator. *Adv. Mater.* **2022**, *34*, 2204508. [CrossRef] [PubMed]
25. Sivarenjini, T.M.; Panbude, A.; Sathiyamoorthy, S.; Kumar, R.; Maaza, M.; Kaliappan, J.; Veluswamy, P. Design and Optimization of Flexible Thermoelectric Coolers for Wearable Applications. *ECS J. Solid State Sci. Technol.* **2021**, *10*, 081006. [CrossRef]
26. Dabrowska, A.; Kobus, M.; Starzak, L.; Pekoslawski, B. Analysis of Efficiency of Thermoelectric Personal Cooling System Based on Utility Tests. *Materials* **2022**, *15*, 1115. [CrossRef] [PubMed]
27. Pop, E.; Sinha, S.; Goodson, K. Heat Generation and Transport in Nanometer-Scale Transistors. *Proc. IEEE* **2006**, *94*, 1587–1601. [CrossRef]
28. Cai, Y.; Wang, Y.; Liu, D.; Zhao, F.Y. Thermoelectric cooling technology applied in the field of electronic devices: Updated review on the parametric investigations and model developments. *Appl. Therm. Eng.* **2019**, *148*, 238–255. [CrossRef]
29. Sharp, J.; Bierschenk, J.; Lyon, H. Overview of Solid-State Thermoelectric Refrigerators and Possible Applications to On-Chip Thermal Management. *Proc. IEEE* **2006**, *94*, 1602–1612. [CrossRef]
30. Adams, M.; Verosky, M.; Zebarjadi, M.; Heremans, J. Active Peltier Coolers Based on Correlated and Magnon-Drag Metals. *Phys. Rev. Applied* **2019**, *11*, 054008. [CrossRef]
31. Parker, M. Going with the flow (of heat). *Nat. Electron.* **2019**, *2*, 211. [CrossRef]
32. Mao, J.; Chen, G.; Ren, Z. Thermoelectric cooling materials. *Nat. Mater.* **2021**, *20*, 454–461. [CrossRef] [PubMed]
33. Zebarjadi, M. Electronic cooling using thermoelectric devices. *Appl. Phys. Lett.* **2015**, *106*, 203506. [CrossRef]
34. Rowe, D.M.; Kuznetsov, V.L.; Kuznetsova, L.A.; Min, G. Electrical and thermal transport properties of intermediate-valence YbAl3. *J. Phys. Appl. Phys.* **2002**, *35*, 2183–2186. [CrossRef]
35. Vandaele, K.; Watzman, S.J.; Flebus, B.; Prakash, A.; Zheng, Y.; Boona, S.R.; Heremans, J.P. Thermal spin transport and energy conversion. *Mater. Today Phys.* **2017**, *1*, 39–49. [CrossRef]
36. Fert, A.; Piraux, L. Magnetic nanowires. *J. Magn. Magn. Mater.* **1999**, *200*, 338–358. [CrossRef]
37. Staňo, M.; Fruchart, O.; Brück, E. Magnetic Nanowires and Nanotubes. In *Handbook of Magnetic Materials*; Elsevier: Amsterdam, The Netherlands, 2018; Volume 27, Chapter 3, pp. 155–267. [CrossRef]
38. He, H.; Tao, N.J. Electrochemical fabrication of metal nanowires. *Encycl. Nanosci. Nanotechnol.* **2003**, *2*, 755–772.
39. Caballero-Calero, O.; Martín-González, M. Thermoelectric nanowires: A brief prospective. *Scr. Mater.* **2016**, *111*, 54–57. [CrossRef]
40. Domínguez-Adame, F.; Martín-González, M.; Sánchez, D.; Cantarero, A. Nanowires: A route to efficient thermoelectric devices. *Phys. Low-Dimens. Syst. Nanostruct.* **2019**, *113*, 213–225. [CrossRef]
41. da Câmara Santa Clara Gomes, T.; Abreu Araujo, F.; Piraux, L. Making flexible spin caloritronic devices with interconnected nanowire networks. *Sci. Adv.* **2019**, *5*, eaav2782. [CrossRef]
42. Abreu Araujo, F.; da Câmara Santa Clara Gomes, T.; Piraux, L. Magnetic Control of Flexible Thermoelectric Devices Based on Macroscopic 3D Interconnected Nanowire Networks. *Adv. Electron. Mater.* **2019**, *5*, 1800819. [CrossRef]
43. da Câmara Santa Clara Gomes, T.; Marchal, N.; Abreu Araujo, F.; Piraux, L. Spin Caloritronics in 3D Interconnected Nanowire Networks. *Nanomaterials* **2020**, *10*, 2092. [CrossRef] [PubMed]
44. Piraux, L.; da Câmara Santa Clara Gomes, T.; Abreu Araujo, F.; De La Torre Medina, J. 3D magnetic nanowire networks. In *Magnetic Nano- and Microwires*, 2nd ed.; Vázquez, M., Ed.; Elsevier: Amsterdam, The Netherlands, 2020; Chapter 27.
45. da Câmara Santa Clara Gomes, T.; Marchal, N.; Araujo, F.A.; Piraux, L. Flexible thermoelectric films based on interconnected magnetic nanowire networks. *J. Phys. Appl. Phys.* **2022**, *55*, 223001. [CrossRef]
46. da Câmara Santa Clara Gomes, T.; Marchal, N.; Abreu Araujo, F.; Piraux, L. Magnetically Activated Flexible Thermoelectric Switches Based on Interconnected Nanowire Networks. *Adv. Mater. Technol.* **2022**, *7*, 2101043. [CrossRef]
47. Ashcroft, N.W.; Mermin, N.D. *Solid State Physics*; Holt, Rinehart and Winston: New York, NY, USA, 1976.
48. Tian, M.; Wang, J.; Kurtz, J.; Mallouk, T.E.; Chan, M.H.W. Electrochemical Growth of Single-Crystal Metal Nanowires via a Two-Dimensional Nucleation and Growth Mechanism. *Nano Lett.* **2003**, *3*, 919–923. [CrossRef] [PubMed]

49. Durkan, C.; Welland, M.E. Size effects in the electrical resistivity of polycrystalline nanowires. *Phys. Rev. B* **2000**, *61*, 14215–14218. [CrossRef]
50. Ye, S.; Rathmell, A.R.; Chen, Z.; Stewart, I.E.; Wiley, B.J. Metal Nanowire Networks: The Next Generation of Transparent Conductors. *Adv. Mater.* **2014**, *26*, 6670–6687. [CrossRef] [PubMed]
51. Yoo, E.; Moon, J.H.; Jeon, Y.S.; Kim, Y.; Ahn, J.P.; Kim, Y.K. Electrical resistivity and microstructural evolution of electrodeposited Co and Co-W nanowires. *Mater. Charact.* **2020**, *166*, 110451. [CrossRef]
52. Meaden, G.T. *Electrical Resistance of Metals*; Springer: New York, NY, USA, 1965.
53. Rowe, D.M. *CRC Handbook of Thermoelectrics*; CRC Press: Boca Raton, FL, USA, 1995.
54. Heremans, J.P.; Wiendlocha, B. Tetradymites: Bi2Te3-Related Materials. In *Materials Aspect of Thermoelectricity*; CRC Press: Boca Raton, FL, USA, 2016; pp. 53–108.
55. Heikes, R.R.; Ure, R.W. *Thermoelectricity: Science and Engineering*; Interscience Publishers: New York, NY, USA, 1961.
56. Zhang, Z.; Chen, J. Thermal conductivity of nanowires. *Chin. Phys. B* **2018**, *27*, 035101. [CrossRef]
57. Yang, X.; Wang, C.; Lu, R.; Shen, Y.; Zhao, H.; Li, J.; Li, R.; Zhang, L.; Chen, H.; Zhang, T.; et al. Progress in measurement of thermoelectric properties of micro/nano thermoelectric materials: A critical review. *Nano Energy* **2022**, *101*, 107553. [CrossRef]
58. Lu, R.; Yang, X.; Wang, C.; Shen, Y.; Zhang, T.; Zheng, X.; Chen, H. Integrated measurement of thermoelectric properties for filamentary materials using a modified hot wire method. *Rev. Sci. Instruments* **2022**, *93*, 125107. [CrossRef]
59. Rojo, M.M.; Calero, O.C.; Lopeandia, A.F.; Rodriguez-Viejo, J.; Martín-Gonzalez, M. Review on measurement techniques of transport properties of nanowires. *Nanoscale* **2013**, *5*, 11526–11544. [CrossRef] [PubMed]
60. Ou, M.N.; Yang, T.J.; Harutyunyan, S.R.; Chen, Y.Y.; Chen, C.D.; Lai, S.J. Electrical and thermal transport in single nickel nanowire. *Appl. Phys. Lett.* **2008**, *92*, 063101. [CrossRef]
61. Kojda, D.; Mitdank, R.; Handwerg, M.; Mogilatenko, A.; Albrecht, M.; Wang, Z.; Ruhhammer, J.; Kroener, M.; Woias, P.; Fischer, S.F. Temperature-dependent thermoelectric properties of individual silver nanowires. *Phys. Rev. B* **2015**, *91*, 024302. [CrossRef]
62. Wang, J.; Wu, Z.; Mao, C.; Zhao, Y.; Yang, J.; Chen, Y. Effect of Electrical Contact Resistance on Measurement of Thermal Conductivity and Wiedemann-Franz Law for Individual Metallic Nanowires. *Sci. Rep.* **2018**, *8*, 4862. [CrossRef]
63. Petsagkourakis, I.; Tybrandt, K.; Crispin, X.; Ohkubo, I.; Satoh, N.; Mori, T. Thermoelectric materials and applications for energy harvesting power generation. *Sci. Technol. Adv. Mater.* **2018**, *19*, 836–862. [CrossRef] [PubMed]
64. Macia, E. *Thermoelectric Materials: Advances and Applications*; CRC Press: Boca Raton, FL, USA, 2015.
65. Yamashita, O.; Tomiyoshi, S.; Makita, K. Bismuth telluride compounds with high thermoelectric figures of merit. *J. Appl. Phys.* **2002**, *93*, 368–374. [CrossRef]
66. Heremans, J.P.; Cava, R.J.; Samarth, N. Tetradymites as thermoelectrics and topological insulators. *Nat. Rev. Mater.* **2017**, *2*, 17049. [CrossRef]
67. Chung, D.Y.; Hogan, T.; Brazis, P.; Rocci-Lane, M.; Kannewurf, C.; Bastea, M.; Uher, C.; Kanatzidis, M.G. $CsBi_4Te_6$: A High-Performance Thermoelectric Material for Low-Temperature Applications. *Science* **2000**, *287*, 1024–1027. [CrossRef]
68. Mao, J.; Zhu, H.; Ding, Z.; Liu, Z.; Gamage, G.A.; Chen, G.; Ren, Z. High thermoelectric cooling performance of n-type Mg_3Bi_2-based materials. *Science* **2019**, *365*, 495–498. [CrossRef]
69. Pan, Y.; Yao, M.; Hong, X.; Zhu, Y.; Fan, F.; Imasato, K.; He, Y.; Hess, C.; Fink, J.; Yang, J.; et al. $Mg_3(Bi,Sb)_2$ single crystals towards high thermoelectric performance. *Energy Environ. Sci.* **2020**, *13*, 1717–1724. [CrossRef]
70. Sun, P.; Ikeno, T.; Mizushima, T.; Isikawa, Y. Simultaneously optimizing the interdependent thermoelectric parameters in $Ce(Ni_{1-x}Cu_x)_2Al_3$. *Phys. Rev. B* **2009**, *80*, 193105. [CrossRef]
71. Boona, S.R.; Morelli, D.T. Enhanced thermoelectric properties of $CePd_{3-x}Pt_x$. *Appl. Phys. Lett.* **2012**, *101*, 101909. [CrossRef]
72. Issi, J.P. Low Temperature Transport Properties of the Group V Semimetals. *Aust. J. Phys.* **1979**, *32*, 585–628. [CrossRef]
73. Blatt, F.J. Magnetic Field Dependence of the Thermoelectric Power of Iron. *Can. J. Phys.* **1972**, *50*, 2836–2839. [CrossRef]
74. Arajs, S.; Anderson, E.E.; Ebert, E.E. Absolute thermoelectric power of chromium-silicon alloys. *Il Nuovo C. B 1971–1996* **1971**, *4*, 40–50. [CrossRef]
75. Ho, C.Y.; Bogaard, R.H.; Chi, T.C.; Havill, T.N.; James, H.M. Thermoelectric power of selected metals and binary alloy systems. *Thermochimica Acta* **1993**, *218*, 29–56. [CrossRef]
76. Wehmeyer, G.; Yabuki, T.; Monachon, C.; Wu, J.; Dames, C. Thermal diodes, regulators, and switches: Physical mechanisms and potential applications. *Appl. Phys. Rev.* **2017**, *4*, 041304. [CrossRef]
77. Klinar, K.; Kitanovski, A. Thermal control elements for caloric energy conversion. *Renew. Sustain. Energy Rev.* **2020**, *118*, 109571. [CrossRef]
78. da Câmara Santa Clara Gomes, T.; de la Torre Medina, J.; Velázquez-Galván, Y.G.; Martínez-Huerta, J.M.; Encinas, A.; Piraux, L. Interplay between the magnetic and magneto-transport properties of 3D interconnected nanowire networks. *J. Appl. Phys.* **2016**, *120*, 043904. [CrossRef]
79. da Câmara Santa Clara Gomes, T.; De La Torre Medina, J.; Lemaitre, M.; Piraux, L. Magnetic and Magnetoresistive Properties of 3D Interconnected NiCo Nanowire Networks. *Nanoscale Res. Lett.* **2016**, *11*, 466. [CrossRef] [PubMed]
80. Ferain, E.; Legras, R. Track-etch templates designed for micro- and nanofabrication. *Nucl. Instrum. Methods Phys. Res. Sect. Beam Interact. Mater. Atoms* **2003**, *208*, 115–122. [CrossRef]
81. Kamalakar, M.V.; Raychaudhuri, A.K. Low temperature electrical transport in ferromagnetic Ni nanowires. *Phys. Rev. B* **2009**, *79*, 205417. [CrossRef]

82. Marchal, N.; da Câmara Santa Clara Gomes, T.; Abreu Araujo, F.; Piraux, L. Large spin-dependent thermoelectric effects in NiFe-based interconnected nanowire networks. *Nanoscale Res. Lett.* **2020**, *15*, 137. [CrossRef] [PubMed]
83. Laubitz, M.J.; Matsumura, T. Transport properties of the ferromagnetic metals. I. Cobalt. *Can. J. Phys.* **1973**, *51*, 1247–1256. [CrossRef]

Disclaimer/Publisher's Note: The statements, opinions and data contained in all publications are solely those of the individual author(s) and contributor(s) and not of MDPI and/or the editor(s). MDPI and/or the editor(s) disclaim responsibility for any injury to people or property resulting from any ideas, methods, instructions or products referred to in the content.

Article

Thermoelectric Power Generation of TiS$_2$/Organic Hybrid Superlattices Below Room Temperature

Numan Salah [1,2,*], Neazar Baghdadi [1,2], Shittu Abdullahi [3,4], Ahmed Alshahrie [1,3] and Kunihito Koumoto [1,5]

1. Center of Nanotechnology, King Abdulaziz University, Jeddah 21589, Saudi Arabia
2. K. A. CARE Energy Research and Innovation Center, King Abdulaziz University, Jeddah 21589, Saudi Arabia
3. Department of Physics, Faculty of Science, King Abdulaziz University, Jeddah 21589, Saudi Arabia
4. Department of Physics, Faculty of Science, Gombe State University, P.M.B, Gombe 127, Nigeria
5. Nagoya Industrial Science Research Institute, Nagoya 464-0819, Japan
* Correspondence: nsalah@kau.edu.sa

Abstract: Recently, the n-type TiS$_2$/organic hybrid superlattice (TOS) was found to have efficient thermoelectric (TE) properties above and near room temperature (RT). However, its TE performance and power generation at the temperature gradient below RT have not yet been reported. In this work, the TE performance and power generation of the TOS above and below RT were investigated. The electrical conductivity (σ) and Seebeck coefficient (S) were recorded as a function of temperature within the range 233–323 K. The generated power at temperature gradients above (at $\Delta T = 20$ and 40 K) and below (at $\Delta T = -20$ and -40 K) RT was measured. The recorded σ decreased by heating the TOS, while $|S|$ increased. The resulting power factor recorded ~100 μW/mK^2 at T = 233 K with a slight increase following heating. The charge carrier density and Hall mobility of the TOS showed opposite trends. The first factor significantly decreased after heating, while the second one increased. The TE-generated power of a single small module made of the TOS at $\Delta T = 20$ and 40 K recorded 10 and 45 nW, respectively. Surprisingly, the generated power below RT is several times higher than that generated above RT. It reached 140 and 350 nW at $\Delta T = -20$ and -40 K, respectively. These remarkable results indicate that TOS might be appropriate for generating TE power in cold environments below RT. Similar TE performances were recorded from both TOS films deposited on solid glass and flexible polymer, indicating TOS pertinence for flexible TE devices.

Keywords: TiS$_2$/organic superlattice; thermoelectric; TE power generation; ΔT below RT; TE module

Citation: Salah, N.; Baghdadi, N.; Abdullahi, S.; Alshahrie, A.; Koumoto, K. Thermoelectric Power Generation of TiS$_2$/Organic Hybrid Superlattices Below Room Temperature. *Nanomaterials* 2023, 13, 781. https://doi.org/10.3390/nano13040781

Academic Editor: Byunghoon Kim

Received: 10 January 2023
Revised: 2 February 2023
Accepted: 7 February 2023
Published: 20 February 2023

Copyright: © 2023 by the authors. Licensee MDPI, Basel, Switzerland. This article is an open access article distributed under the terms and conditions of the Creative Commons Attribution (CC BY) license (https://creativecommons.org/licenses/by/4.0/).

1. Introduction

Global energy demands increase by the day in all sectors of society. This is due to the extensive use of electricity in modern electronic devices and facilities, various mobilities, industries, and infrastructures. However, this has resulted in serious environmental pollution and a climate crisis, particularly due to the use of fossil fuels. As these sources are limited and non-sustainable, it is of great importance to find other energy sources and alternative technologies. One of the promising technologies is converting solar energy or waste heat from various sources into electrical energy using effective thermoelectric (TE) materials. There is a large number of TE materials that have been developed and attempts have been made to enhance/improve these for better TE performance. The first developed TE material was Bi$_2$Te$_3$ [1], this was then followed by n-type Bi$_2$Te$_{3-x}$Se$_x$ and p-type Bi$_{2-x}$Sb$_x$Te$_3$ [2]. Subsequently, a wide variety of TE materials, such as sulfides, selenides, silicides, skutterudites, intermetallic compounds, oxides, organic polymers, and carbon nanomaterials, etc. [3–11] have been developed. These were mostly developed as mid- to high-temperature TE materials, some of which include SnSe, SnS$_{0.91}$Se$_{0.09}$, SnTe, GeTe, and Cu$_2$Te-based compounds, etc. [12–18] However, low-temperature TE materials are still rarely developed. Moreover, it is noticed that most of the developed materials were investigated for their power generation at temperature gradients only far above RT.

Thus far, the best low-temperature TE materials are Bi_2Te_3-based compounds [19]. However, the generated power using these compounds at low temperatures is rarely reported in the literature. Moreover, improving its intrinsic poor mechanical properties and lowering the content of toxic tellurium is still a big challenge for extensive applications. Afterward, some efforts were made to develop other low-temperature TE materials such as BiSb alloys [20], Ta_4SiTe_4 crystal in its one-dimensional form [21], $Ce(Ni_{1-x}Cu_x)_2Si_2$ and $CeNi_2(Si_{1-y}Ge_y)_2$ [22], CoSi and $Co_{1-x}M_xSi$ [23]. Recent reports on Ag_2Se showed that this material is a promising low-temperature TE material due to its high ZT value, intrinsic semiconductor nature, ultra-high carrier mobility, small density-of-states effective mass, and ultra-low lattice thermal conductivity [24,25]. Mg_3Bi_2 [26], Ta_4SiTe_4 [27], GeTe-based alloys [28], and silicon thin films [29] were also reported to have moderately good TE performance at low temperatures. Some of these materials seem to be promising; however, the effect of incorporating them with some polymers to produce flexible TE devices is still unknown and their power generation at temperature gradients below RT has not been reported.

For flexible TE materials and devices, highly anisotropic crystals such as $CsBi_4Te_6$ or TiS_2 might be useful for intercalation with some organic thin layers. Recent reports showed that $CsBi_4Te_6$ has a good TE performance at low temperatures [30]. In addition, TiS_2 has an anisotropic layered structure with attractive electrical properties. Furthermore, $1T$-TiS_2 has a hexagonal close-packed (hcp) structure, with its layers consisting of covalent Ti-S bonds with a bond length of 2.423 Å. Such layers are bound together by the van der Waals force. Moreover, due to its elegant features such as an attractive structure, lightweight material, cheap chalcogenide, and high electrical conductivity with a semi-metallic behavior [31], TiS_2 has been employed as a cathode material in rechargeable batteries [32], electrode materials of pseudocapacitors [33], and as a sensor material for uric acid determination [34]. In addition to these applications, the single crystals of TiS_2 were reported to show good TE performance [35–38]. It was also involved in nanocomposites [39] for producing flexible TE materials and devices [40–42]. Additionally, layered TiS_2 intercalated with linear alkylamines has recently attracted significant interest as a model compound for flexible n-type thermoelectric applications, showing remarkably high-power factors at RT [43]. The excellent anisotropic property of TiS_2 led to the novel work on developing a superlattice made of TiS_2/organic, which seems to be very important from the application point of view. However, the TE performance of the TiS_2/organic superlattice and its power-generation characteristics below RT have not been reported yet in the literature.

In this work, the n-type TiS_2/organic hybrid superlattice (TOS) was produced and investigated for its TE performance below RT. Moreover, the power generation at a temperature gradient below RT was measured. The electrical conductivity (σ) and Seebeck coefficient (S) were recorded as a function of temperature within the range 233–323 K. Additionally, the power-generation characteristics at the temperature gradients $\Delta T = 20$ K and 40 K were measured either below or above RT. The charge carrier density and Hall mobility of TOS were measured at 233–323 K. The generated maximum power of a small module made of TOS (formed on a glass substrate) at $\Delta T = 20$ and 40 K above RT were compared with those generated below RT. The obtained results are discussed in detail and showed the significant potential of this material to be used for generating TE power in cold environments.

2. Materials and Methods

Synthesis of the TiS_2/organic superlattice thin film was conducted in this work, similar to that reported by Tian et al. [40]. Commercially available TiS_2 powder of high purity (99.9%) with a particle size of around 200 mesh was obtained from Sigma Aldrich, Bengaluru, India. While hexylamine, HA (99.0%,) and N, N-dimethylformamide, DMF (99.0%) were supplied from Sigma Aldrich, Darmstadt, Germany. Initially, TiS_2 powder was mixed and ground with hexylamine (HA) in an agate mortar with a molar ratio of 1:4, and this resulted in a metallic brown powder. This resultant powder was further exfoliated into nanosheets by adding DMF as a highly polar solvent. A volume ~20 times that of the HA was added. This addition was followed by sonication. The obtained colloidal solution was then centrifuged to remove the unintercalated TiS2 and thick flakes. The exfoliated nanosheets were added into a petri dish containing a glass slide to make thin films on. The petri dish was inserted inside a vacuum furnace at 60 °C for drying. The obtained films formed on the glass slide had a thickness of 40 µm. The film was further annealed at 403 K for 5 h. A polymer substrate was also used to produce a flexible TE material. This polymer is a polyethylene terephthalate (PET) sheet of a thickness of around 20 µm. It was supplied by Fuji Film Holdings, Japan.

The films were characterized using several techniques, such as scanning electron microscopy (SEM) (JSM-7500F, JEOL, Tokoyo, Japan), transmission electron microscopy (TEM) (JEM 2100F, JEOL), and X-ray diffraction system (XRD) (ULTIMA IV, Rigaku, Japan). The electrical conductivity, Seebeck coefficient, and thermal conductivity in the in-plane directions were recorded for the prepared film as a function of temperature. An LSR-3 Linseis-Seebeck coefficient and electric resistivity system manufactured by Linseis, Germany was used in a helium atmosphere to measure the films' resistivities and Seebeck coefficients. The heating rate and temperature gradient between the hot and cold sides were fixed at 5 K/min and 50 K, respectively. The charge carrier concentration and Hall mobility were measured using the HCS 10 system, Linseis. A single leg of the TE module made of a film of dimensions: thickness × length × width = 0.03 × 8 × 8 mm was constructed. The in-plane thermal conductivity of the fabricated TOS was determined using the laser flash method in LFA-1000 (Linseis, Selb, Germany). The measurements were performed using a sample holder made of graphite supplied by Linseis. The measurement was conducted in a vacuum atmosphere and the heating rate was set at 10 K/min. The TE leg was perpendicularly fixed using a stand made of a ceramic plate, while aluminum electrodes were used to attach both sides of the TE leg to the measurement systems. To control the temperature gradient between two sides of the leg, a hot plate, solid ice, cold water, and liquid nitrogen were employed to generate the temperature differences (a table and figure describing this arrangement are included in the following sections). An infrared temperature gun to measure the temperature at the edge of the leg was used. The power-generation characteristics of this single leg were investigated at different temperatures under real-time conditions in the air.

3. Results and Discussion

The surface morphology and microstructure of the produced TiS_2/organic superlattice (TOS) film on a glass substrate were investigated by both SEM and TEM techniques and the obtained result is presented in Figures 1 and 2. The SEM images shown in Figure 1a,b clearly display the aligned layers/flakes of TOS with a thickness of less than 100 nm. Some gaps can be seen between these layers. These layers/flakes of TOS seem not to be completely stacked with each other, at least in those which are close to the edges. Nevertheless, this structure and these layers are similar to those reported previously [40–45]. The film color is also like those reported in the literature, as shown in the photographed image presented in Figure 1c, which shows a top view of the film formed on a 10 × 10 mm glass substrate. HRTEM images of a TOS film were also recorded at two different magnifications and are presented in Figure 2a,b. These images show the atomic and sub-atomic details of a single layer/flake of TOS, which firmly indicate that a nearly perfect stacking of a TiS2/organic layer-by-layer structure was achieved. The line scan of the HRTEM image is presented in Figure 2c, which shows the d spacing of the corresponding (001) plane is ~9.64 Å. This value is much larger than that of a TiS_2 single crystal, ~5.4 Å [38], as shown in Figure 2d, which also confirms that a nearly perfect superlattice composed of TiS_2/organic alternate stacking layers was formed [44].

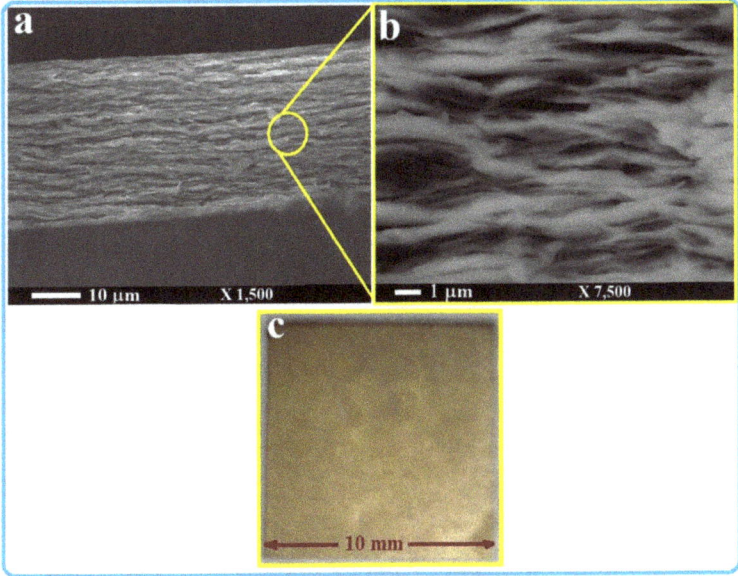

Figure 1. SEM images (**a**,**b**) of the synthesized film of TiS$_2$/organic superlattice. Photograph image of a top view of the film formed on a 10 × 10 mm glass substrate is also shown (**c**).

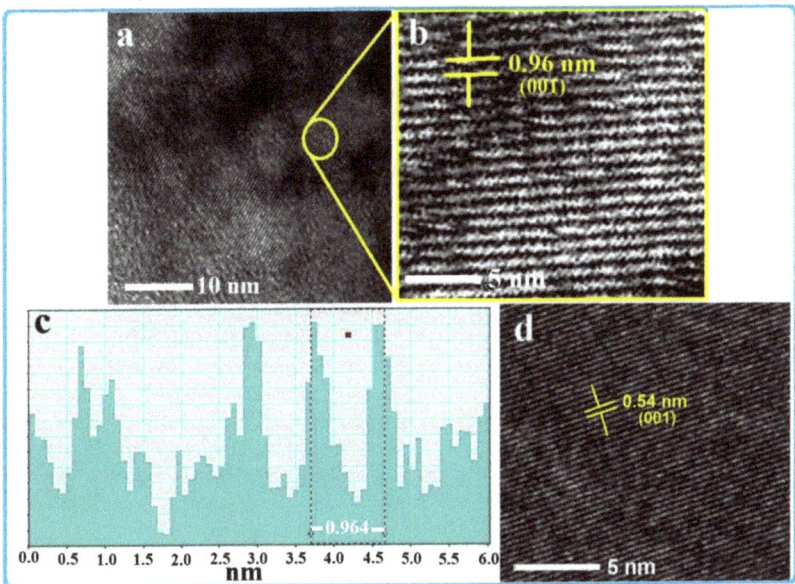

Figure 2. HRTEM images (**a**,**b**) of TiS$_2$/organic superlattice and line scan (**c**) of the HRTEM image showing a d-spacing of ~9.64 Å. The HRTEM image presented in (**d**) is of a TiS$_2$ single crystal.

As mentioned above in the experimental section, a polymer substrate was also used to produce a flexible TE material. This polymer was chosen to be a polyethylene terephthalate (PET) sheet of a thickness of around 20 µm (Figure 3). In Figure 3a, a SEM image of the TOS deposited on the PET substrate is shown. The thickness of the deposited TOS is around 25 µm. The layers shown in Figure 3b are like those deposited on a glass substrate (Figure 1b). The photographs of the PET film without and with TOS are shown in Figure 3c,d, respectively. The deposited TOS film was found to be very stable even with bending the film several times. No crack formation or significant effect on TE performance was observed. Its TE performance and power generation are similar to those of the film deposited on a solid-glass substrate (shown in the next sections). This is quite important to fabricate TE materials for different systems and flexible devices.

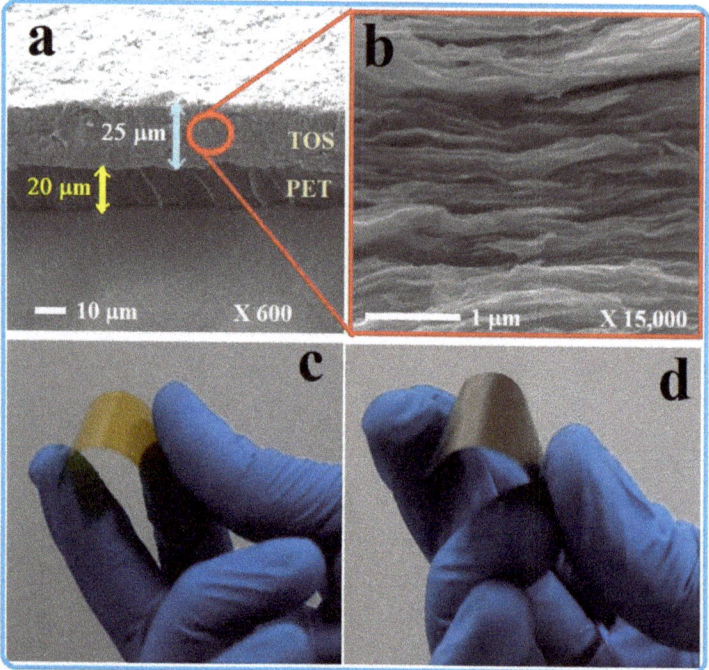

Figure 3. (**a**) SEM image of the TiS$_2$/organic superlattice deposited on a PET substrate. (**b**) High-magnification SEM image of the TOS deposited on the PET substrate. Photographs of the PET film (**c**) without TOS and (**d**) with deposited TOS.

In addition to the above investigations, which described the surface morphology and microstructure of the TOS (Figures 1 and 2), the XRD pattern of the film formed on a glass substrate as well as that of a pressed compact of collected powder (collected without using any substrate) was recorded. These measurements were recorded after annealing these samples at 403 K for 5 h. The obtained results are presented in Figure 4a,b. The XRD pattern of the TOS film displays several peaks. These peaks were reported to be of two intercalated phases due to the different arrangements of organic molecules in the van der Waals gap of TiS$_2$ [40]. The hkl values were labeled in two different colors: blue for phase one and black for the second phase. These two phases were reported to have different lattice parameters c, with values of 17.1 Å and 9.92 Å. In the present samples, the major phase present in both the film and the pellet (Figure 3a,b) is that of the lattice parameter 9.92 Å. The HRTEM image presented in Figure 2b shows a d spacing of 9.64 Å, indicating that this phase is the major one present here. Nevertheless, these results are similar to those produced and evaluated previously [40].

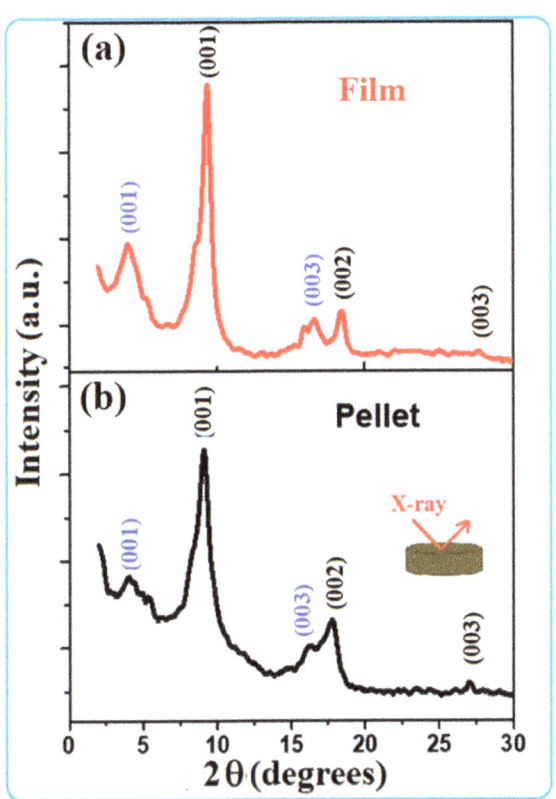

Figure 4. XRD patterns of TOS formed as (**a**) a film on a glass substrate, and (**b**) a pressed compact of the collected powder formed on the bottom of the petri dish without using a substrate.

Seebeck coefficient at 233 K was recorded as −55 μV/K. This value increased to approximately −72 μV/K by heating the film to 323 K. The obtained values are all negative, indicating that this semiconductor has n-type carriers, which is similar to that of TiS_2 reported by many workers [35–38]. The calculated power factor, *PF*, is shown in Figure 5b. The obtained value at 233 K is ~104 μW/mK^2. This value is found to slightly increase with heating. It recorded around 120 μW/mK^2 at 323 K. This behavior proved to be of almost a degenerate semiconductor, confirming its suitability as an n-type TE material.

The charge carrier density, Hall mobility, charge carrier effective mass, and charge carrier mean free path of the TOS film were also recorded as a function of temperature and the obtained results are presented in Figure 6. It is clear from Figure 5a that the charge carrier density and Hall mobility curves have opposite trends against temperature. The recorded carrier density at 233 K is ~0.75 × 10^{21} cm^{-3}, but it increased by a factor of three with heating up to 323 K. The Hall mobility is ~2.8 ± 0.5 cm^2/Vs at 233 K but decreased to approximately 0.65 ± 0.11 cm^2/Vs by heating up to 323 K. The obtained charge carrier density and mobility values in this study are close to those reported in the literature [40,46]. However, the observed behavior of increasing the charge concentration by heating might be due to the generation of more carriers and their participation in the electrical conduction, while the decrease in their mobility might have occurred as a result of lattice expansion, which would affect the available carrier channels.

To further investigate the electrical conductivity, the effective mass (*m**) and mean free path of the free charge carriers in TOS as a function of temperature were calculated and plotted in Figure 6b. It is clear from the above result that the electrical conductivity

(Figure 5a) of the TOS film seems to be of a degenerate semiconductor; therefore, the effective mass can be obtained from the Seebeck coefficient, S, and the carrier concentration, n, according to Pisarenko's relationship [47]:

$$S = \frac{8\pi^2 k_B^2}{3eh^2} m^* T \left(\frac{\pi}{3n}\right)^{2/3} \quad (1)$$

where k_B is Boltzmann constant = 1.38 × 10^{-23} m^2 kgs^{-2} K^{-1}, h is Planck's constant= 6.63 × 10^{-34} m^{2}kg/s, and e = the electron charge = 1.6 × 10^{-19} Coulomb.

The electron mean free path l was deduced from the Fermi velocity v_F and the scattering time (or relaxation time) τ [46]:

$$l = v_F \tau \quad (2)$$

The relaxation time (τ) can be calculated using the following equation:

$$\tau = \frac{\sigma m^*}{ne^2} \quad (3)$$

The Fermi velocity at the Fermi surface was calculated using the following equation [48]:

$$v_F = \frac{\hbar k_F}{m_0} = \frac{\hbar}{m_0}(3\pi^2 n)^{\frac{1}{3}} \quad (4)$$

where $\hbar$ = reduced Planck constant, m_0 = electron mass, and k_F = the Fermi wavevector.

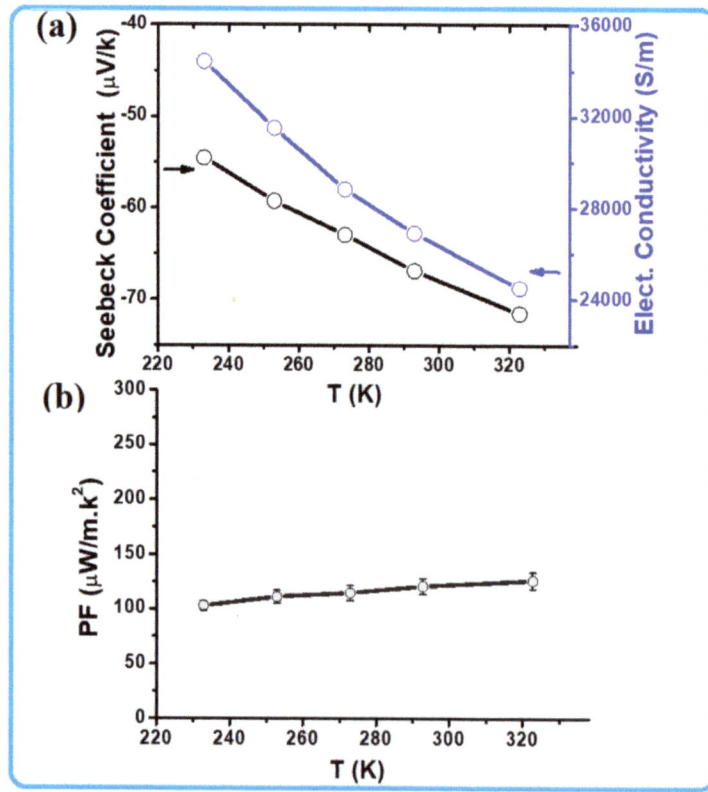

Figure 5. TE performance of the TOS film as a function of temperature.

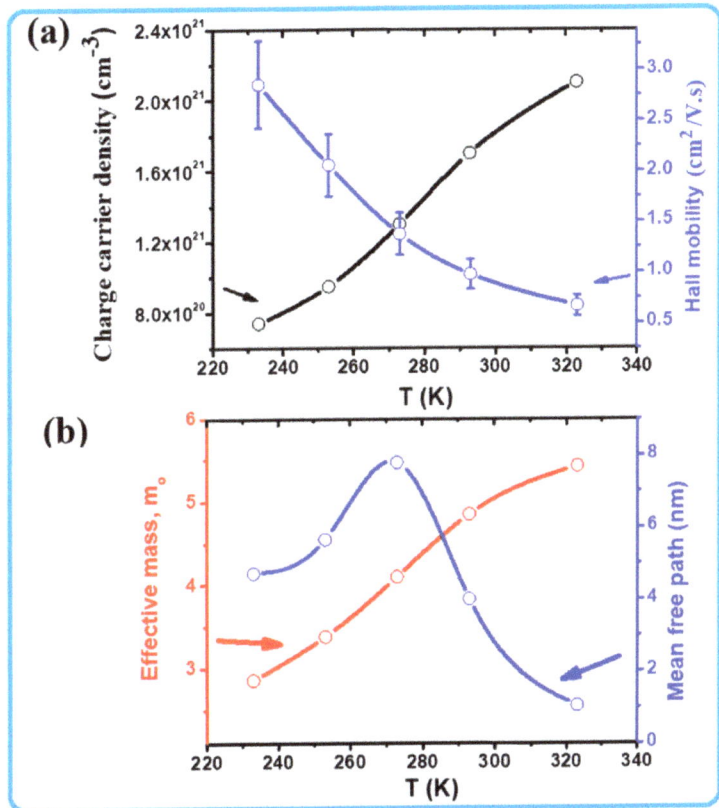

Figure 6. (a) Charge carrier density, Hall mobility, and (b) effective mass and mean free path of the free charge carriers in TOS as a function of temperature.

The effective mass values were found to increase with heating from around 2.8 m_0 at 233 K to approximately 5.5 m_0 at 323 K (Figure 6b). This high value and its increase with increasing temperature is the main reason for the obtained Seebeck coefficient, whose absolute value also increased with heating. In the case of the mean free path, the maximum value, 8 nm, was detected at ~273 K. Below 273 K, the mean free path reduced to 5 nm, while above 273 K, this value significantly decreased, to 1 nm at 323 K. The mean free path of 8 nm seems to be close to the interlayer distance of TOS, but by changing temperature, the interlayer distance might have decreased due to either layer expansion or shrinkage. However, it is difficult to model the relationship between the carrier mean free path and lattice distortion; thus, this phenomenon should be subject to further investigation.

The thermal conductivity (κ_{total}) of a material is the sum of electronic thermal conductivity (κ_e) and lattice (phonon) thermal conductivity (κ_p). The value of κ_e can be obtained using the following Wiedemann–Franz law [49] after measuring the electrical conductivity.

$$L = \frac{\kappa_e}{\sigma T} = \frac{\pi^2 k_B^2}{3e^2} = 2 \times 10^{-8} \, W\Omega K^{-2} \tag{5}$$

where L is the Lorenz number, k_B is Boltzmann's constant, and e is the electron charge. The values of the recorded in-plane κ_{total}, κ_p and κ_e of the present TOS are presented as a function of temperature in Table 1. They were recorded within the range 298–363 K. At 298 K, the κ_{total} of the TOS was equal to 0.76 W/mK. When the temperature was increased to 363 K, this value increased to around 0.97 W/mK. The κ_{total} seems to be temperature-dependent within this temperature range. Nevertheless, these values are comparable to those published in the literature [41]. As the temperature increases, the TOS layers may expand, resulting in close contact between them. Consequently, it may reduce phonon scattering sites and enhance phonon transport, which would lead to increased thermal conductivity, κ_p. The κ_e of the TOS is small compared to its κ_{total}, while the κ_p is closer to its κ_{total}, as displayed in Table 2. This indicates that the phonons are the major heat carriers in the TOS. At temperatures below 298 K the values of κ_{total} might be much smaller; this will be considered in our future work. The assessment of the figure of merit (ZT) as a function of temperature within the range 298–363 K of the present TOS is shown in Table 1. ZT is almost independent of temperature within this temperature range, while PF increased with increasing temperature.

Table 1. The in-plane total thermal conductivity, κ_{total}, phonon thermal conductivity, κ_p, electronic thermal conductivity, κ_e, and figure of merit, ZT of the TiS$_2$/organic superlattice film as a function of temperature within the range 298–363 K. The values of the density; specific heat capacity, C; thermal diffusivity, D; electrical conductivity, σ; and power factor, PF of the TOS film are also shown in this temperature range.

T (K)	Density (g/cm^3)	C (J/gK)	D (cm^2/s)	κ_{total} (W/mK)	κ_p (W/mK)	κ_e (W/mK)	σ (S/m)	PF (µW/mK^2)	ZT
298		0.582	0.0136	0.76	0.60	0.160	27,000	120	0.047
323	0.96	0.585	0.0145	0.81	0.65	0.156	24,200	124	0.049
348		0.601	0.0156	0.94	0.792	0.148	21,300	127	0.047
363		0.619	0.0164	0.97	0.840	0.131	18,100	129	0.048

Table 2. The selected temperatures and temperature gradients are used to measure the power-generation characteristics of a single-leg module of TOS film above/near and below RT.

	T_1 (K)	T_2 (K)	$\Delta T = T_2 - T_1$ (K)	Remarks
Above/near RT	283	303	20	A heater was used at T_2, cold water at T_1
	293	333	40	A heater was used at T_2, RT air at T_1
Below RT	273	253	−20	Liquid nitrogen with a thin ceramic plate was used at T_2, solid ice at T_1
	273	233	−40	Liquid nitrogen was used at T_2, solid ice at T_1

The TE power characteristics of a single-leg module of the TOS film above/near RT at two different temperature differences, ΔT, above and below RT were investigated. As mentioned above, a thin film formed on a glass substrate with thickness = 25 µm, length = 8 mm, and width = 8 mm was used. Table 2 summarizes the selected temperatures and temperature gradients, ΔT was used to measure the power generation produced from this single-leg module. Figure 7 shows a schematic diagram of the single-leg module and the used setup to measure the power generation of this TE device above (Figure 7a) and below (Figure 7b) RT. The obtained results were plotted and are presented in Figures 8 and 9, respectively. When ΔT was fixed at 20 K and 40 K above RT, the recorded short-circuit currents were found to be approximately 75 and 165 µA, while the open-circuit voltages reached 0.5 and 1.1 mV, respectively (Figure 8A). The corresponding power values were plotted as a function of current and are shown in Figure 8B. At $\Delta T = 20$ K, the maximum power was recorded around 10 nW, while at $\Delta T = 40$ K, the maximum power reached 45 nW. These values are reasonable for a single leg made of a small foil [40]. The obtained power values were plotted as a function of load and are shown in Figure 8C. The curves in this figure clearly show that the maximum power values were obtained at an external load of approximately 7 Ω for both $\Delta T = 20$ K and 40 K. This indicated that the internal resistance of the used film is close to 7 Ω. This value is low, which is consistent with the value of the electrical conductivity of the present TOS film.

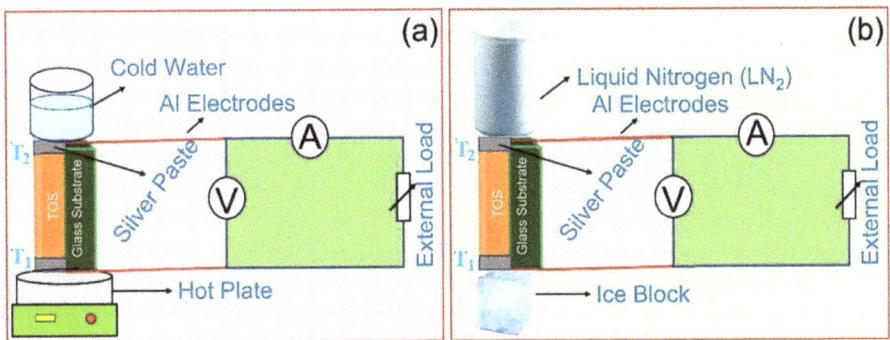

Figure 7. A schematic diagram of the TE device structure and the user setup to measure the power-generation characteristics of a single-leg module of TOS film (**a**) above ($\Delta T = 20$ K and 40 K) and (**b**) below RT ($\Delta T = -20$ K and -40 K).

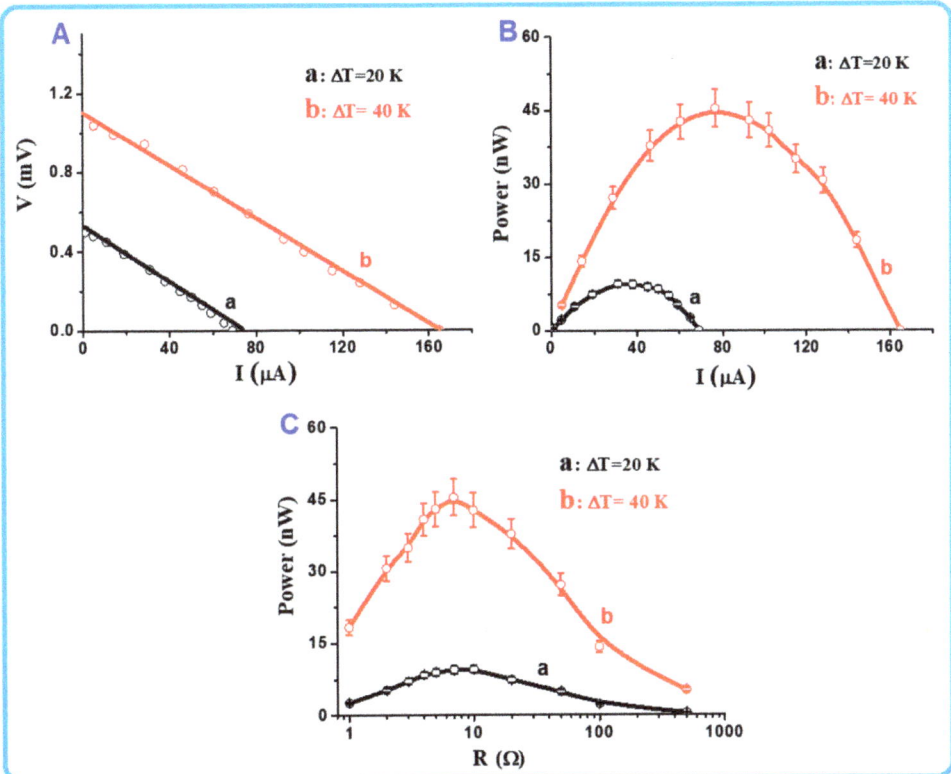

Figure 8. (**A–C**) Above/near RT thermoelectric power characteristics of TOS single leg module at $\Delta T = 20$ K and 40 K (for $\Delta T = 20$ K; $T_1 = 283$ K and $T_2 = 303$ K, for $\Delta T = 40$ K; $T_1 = 293$ K and $T_2 = 333$ K).

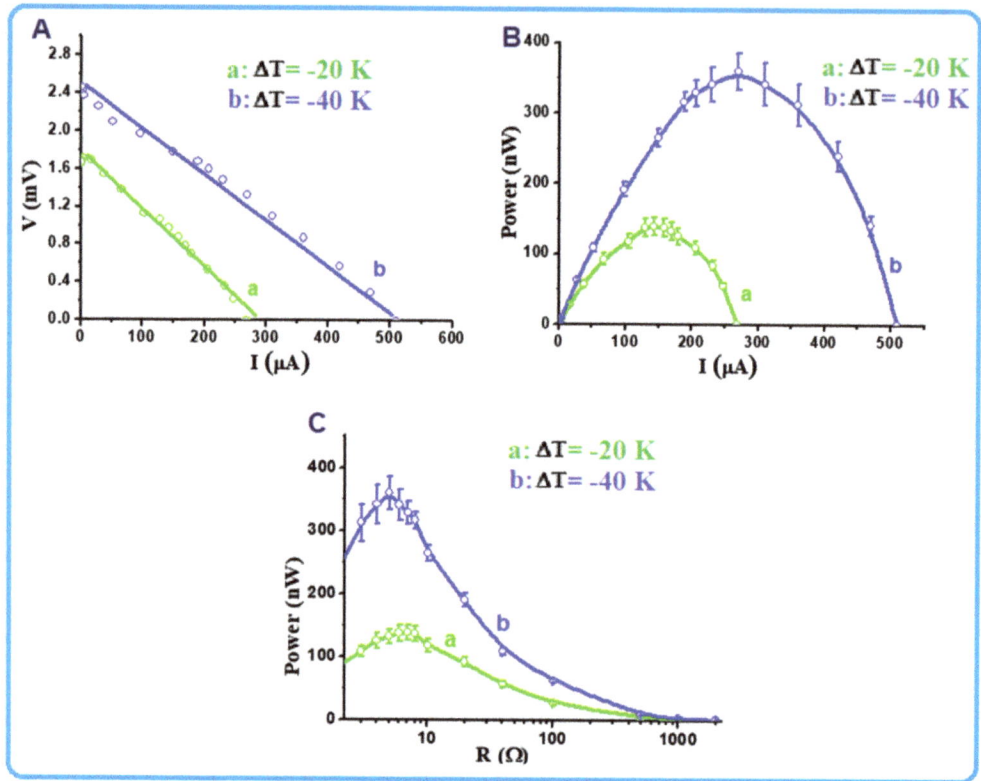

Figure 9. (**A–C**) Below RT thermoelectric power characteristics of the TOS single-leg module at ΔT = −20 K and −40 K (for ΔT = −20 K: T_1 = 273 K and T_2 = 253 K, for ΔT = −40 K: T_1 = 273 K and T_2 = 233 K).

The generated power below RT for the single-leg module made of the TOS film was also measured and the obtained results are shown in Figure 9. Two temperature gradients, ΔT, were selected, −20 K and −40 K, as shown in Table 2. Similarly, when ΔT was fixed at −20 K and −40 K below RT, the recorded short-circuit currents were found to be approximately 280 and 520 µA, while the open-circuit voltages reached 1.7 and 2.5 mV, respectively (Figure 9A). The corresponding power values were plotted as a function of current and are shown in Figure 9B. At ΔT = −20 K, the maximum power recorded was around 140 nW, while at ΔT = −40 K, the maximum power reached 350 nW. Surprisingly, these values were found to be several times higher than those obtained at temperature gradients above RT (Figure 8B). The obtained power values were also plotted as a function of load and the obtained results are shown in Figure 9C. The curves in this figure clearly show that the maximum power values were obtained at an external load of approximately 4–5 Ω for both ΔT = −20 K and −40 K. This indicated that the internal resistance of the used foil is close to this range, e.g., 4–5 Ω. This value is lower than that obtained when the value of ΔT was fixed above RT (Figure 8C). Although the achieved power density of this single small module (area = 64 mm^2) was found to be 5.47 mW/m^2 at ΔT = −40 K, the obtained results can still be further enhanced by changing the polar organic molecules used in the superlattice formation, as reported recently by another research group [50].

The new remarkable finding on TOS in this work is the generation of highly improved TE power below RT, which is several times higher than that generated above/near RT at the same ΔT values. Similarly, here, below RT, the generated current is very high compared to the generated voltage, suggesting that this kind of TE module is suitable as a current source for some devices requiring higher currents. This might be useful in cold-environment regions, particularly in the winter season. The reason behind this improvement below RT might be related to the enhancement of both electrical conductivity and the Seebeck coefficient occurring below RT, as shown in Figure 5a. The reduction in thermal conductivity (hopefully below RT), as shown in Table 2, would keep a comparatively high ZT, giving credit for energy conversion efficiency below RT.

The present remarkable findings would firmly suggest that the TOS films might be useful to generate TE power in cold environments. Moreover, the TOS film can be formed on a solid-glass substrate or a flexible-polymer substrate (Figure 3) and both can show almost the same TE performance. This feature can extend the TOS application as a promising heat-driven power source for a wide range of flexible/wearable electronic systems.

4. Conclusions

In this work, the n-type TiS_2/organic hybrid superlattice (TOS) was found to have efficient TE properties below RT. In particular, a single-leg module made up of a TOS film showed remarkable power-generation characteristics. The generated TE power below RT was found to be more than eight times higher than those generated above RT. It was recorded as 140 and 350 nW at $\Delta T = -20$ K and -40 K, respectively. This might be related to the enhancement of both electrical conductivity and the Seebeck coefficient, while rather high ZT was maintained below RT due to effectively lowered thermal conductivity; however, this phenomenon needs to be further investigated in future work. From the application point of view, this finding would suggest that TE TOS devices might be useful for small-scale TE power generation in cold environments, for which largely extended modules might be designed and fixed to house windows, for instance, to generate TE power from the temperature difference between inside and outside.

Author Contributions: N.S. contributed to the conceptualization, supervision, investigation, and writing original draft; N.B. conducted the TE measurements and contributed to the investigation; S.A. contributed to the data curation and power measurement; A.A. contributed to the funding acquisition, project administration, and investigation; K.K. contributed to the conceptualization, review, and editing. All authors have read and agreed to the published version of the manuscript.

Funding: This project was funded by Science and Technology Unit—King Abdulaziz University—Kingdom of Saudi Arabia—award number (UE-41-104).

Data Availability Statement: The data supporting the findings of this study are available upon request to nsalah@kau.edu.sa.

Conflicts of Interest: The authors declare no conflict of interest.

References

1. Heikes, R.R.; Ure, R.W.; Angello, S.J.; Bauerle, J.E. *Thermoelectricity: Science and Engineering*; Interscience Publishers: New York, NY, USA; London, UK, 1961.
2. Snyder, G.J.; Toberer, E.S. Complex thermoelectric materials. *Nat. Mater.* **2008**, *7*, 105–114. [CrossRef] [PubMed]
3. Zhang, Y.; Li, J.; Hu, W.; Yang, X.; Tang, X.; Tan, G. Boosting thermoelectric performance of SnTe by selective alloying and band tuning. *Mater. Today Energy* **2022**, *25*, 100958. [CrossRef]
4. Li, L.; Kang, W.; Zhao, Y.; Li, Y.; Shi, J.; Cheng, B. Preparation of flexible ultra-fine Al_2O_3 fiber mats via the solution blowing method. *Ceram. Int.* **2015**, *41*, 409–415. [CrossRef]
5. Zhang, W.; Zhu, K.; Liu, J.; Wang, J.; Yan, K.; Liu, P.; Wang, Y. Influence of the phase transformation in Na_xCoO_2 ceramics on thermoelectric properties. *Ceram. Int.* **2018**, *44*, 17251–17257. [CrossRef]
6. Ahmad, A.; Hussain, M.; Zhou, Z.; Liu, R.; Lin, Y.-H.; Nan, C.-W. Thermoelectric performance enhancement of vanadium doped n-Type In_2O_3 ceramics via carrier engineering and phonon suppression. *ACS Appl. Energy Mater.* **2020**, *3*, 1552–1558. [CrossRef]
7. Stevens, D.L.; Parra, A.; Grunlan, J.C. Thermoelectric performance improvement of polymer nanocomposites by selective thermal degradation. *ACS Appl. Energy Mater.* **2019**, *2*, 5975–5982. [CrossRef]
8. Almasoudi, M.; Zoromba, M.S.; Abdel-Aziz, M.H.; Bassyouni, M.; Alshahrie, A.; Abusorrah, A.M.; Salah, N. Optimization preparation of one-dimensional polypyrrole nanotubes for enhanced thermoelectric performance. *Polymer* **2021**, *228*, 123950. [CrossRef]

9. Chakraborty, P.; Ma, T.; Zahiri, A.H.; Cao, L.; Wang, Y. Carbon-based materials for thermoelectrics. *Adv. Cond. Matter Phys.* **2018**, *2018*, 3898479. [CrossRef]
10. Baghdadi, N.; Zoromba, M.S.; Abdel-Aziz, M.H.; Al-Hossainy, A.F.; Bassyouni, M.; Salah, N. One-dimensional nanocomposites based on polypyrrole-carbon nanotubes and their thermoelectric performance. *Polymers* **2021**, *13*, 278. [CrossRef]
11. Famengo, A.; Famengo, A.; Ferrario, A.; Boldrini, S.; Battiston, S.; Fiameni, S.; Pagura, C.; Fabrizio, M. Polyaniline– carbon nanohorn composites as thermoelectric materials. *Polymer Int.* **2017**, *66*, 1725–1730. [CrossRef]
12. He, W.; Wang, D.; Wu, H.; Xiao, Y.; Zhang, Y.; He, D.; Feng, Y.; Hao, Y.J.; Dong, J.F.; Chetty, R.; et al. High thermoelectric performance in low-cost $SnS_{0.91}Se_{0.09}$ crystals. *Science* **2019**, *365*, 1418–1424. [CrossRef] [PubMed]
13. Bu, Z.; Zhang, X.; Shan, B.; Tang, J.; Liu, H.; Chen, Z.; Lin, S.; Li, W.; Pei, Y. Realizing a 14% single-leg thermoelectric efficiency in GeTe alloys. *Sci. Adv.* **2021**, *7*, eabf2738. [CrossRef] [PubMed]
14. Chen, Z.; Sun, Q.; Zhang, F.; Mao, J.; Chen, Y.; Li, M.; Chen, Z.-G.; Ang, R. Mechanical alloying boosted SnTe thermoelectrics. *Mater. Today Phys.* **2021**, *17*, 100340. [CrossRef]
15. Zhao, K.; Liu, K.; Yue, Z.; Wang, Y.; Song, Q.; Li, J.; Guan, M.; Xu, Q.; Qiu, P.; Zhu, H.; et al. Are Cu_2Te-based compounds excellent thermoelectric materials? *Adv. Mater.* **2019**, *31*, 1903480. [CrossRef] [PubMed]
16. Zhao, L.-D.; Lo, S.-H.; Zhang, Y.; Sun, H.; Tan, G.; Uher, C.; Wolverton, C.; Dravid, V.P.; Kanatzidis, M.G. Ultralow thermal conductivity and high thermoelectric figure of merit in SnSe crystals. *Nature* **2014**, *508*, 373–377. [CrossRef]
17. Zhao, L.-D.; Tan, G.; Hao, S.; He, J.; Pei, Y.; Chi, H.; Wang, H.; Gong, S.; Xu, H.; Dravid, V.P.; et al. Ultrahigh power factor and thermoelectric performance in hole-doped single-crystal SnSe. *Science* **2016**, *351*, 141–144. [CrossRef]
18. Qin, B.; Wang, D.; Liu, X.; Qin, Y.; Dong, J.-F.; Luo, J.; Li, J.-W.; Liu, W.; Tan, G.; Tang, X.; et al. Power generation and thermoelectric cooling enabled by momentum and energy multiband alignments. *Science* **2021**, *373*, 556–561. [CrossRef]
19. Zhu, B.; Liu, X.X.; Wang, Q.; Qiu, Y.; Shu, Z.; Guo, Z.T.; Tong, Y.; Cui, J.; Gu, M.; He, J.Q. Realizing record high performance in n-type Bi_2Te_3-based thermoelectric materials. *Energy Environ. Sci.* **2020**, *13*, 2106–2114. [CrossRef]
20. Ibrahim, A.M.; Thompson, A.D. Thermoelectric properties of BiSb alloys. *Mater. Chem. Phys.* **1985**, *12*, 29–36. [CrossRef]
21. Inohara, T.; Okamoto, Y.; Yamakawa, Y.; Yamakage, A.; Takenaka, K. Large thermoelectric power factor at low temperatures in one-dimensional telluride Ta4SiTe4. *Appl. Phys. Lett.* **2017**, *110*, 183901. [CrossRef]
22. Synoradzki, K.; Toliński, T.; Koterlyn, M. Enhanced thermoelectric power factors in the $Ce(Ni_{1-x}Cu_x)_2Si_2$ and $CeNi_2(Si_{1-y}Ge_y)_2$ alloys. *Acta Phys. Pol. A* **2018**, *133*, 366–368. [CrossRef]
23. Ivanov, Y.; Levin, A.A.; Novikov, S.; Pshenay-Severin, D.; Volkov, M.; Zyuzin, A.; Burkov, A.; Nakama, T.; Schnatmann, L.; Reith, H.; et al. Low-temperature thermal conductivity of thermoelectric $Co_{1-x}M_xSi$ (M = Fe, Ni) alloys. *Mater. Today Energy* **2021**, *20*, 100666.
24. Lei, Y.; Qi, R.; Chen, M.; Chen, H.; Xing, C.; Sui, F.; Gu, L.; He, W.; Zhang, Y.; Baba, T.; et al. Microstructurally tailored thin β-Ag2Se films toward commercial flexible thermoelectrics. *Adv. Mater.* **2021**, *34*, 2104786.
25. Jin, M.; Liang, J.; Qiu, P.; Huang, H.; Yue, Z.; Zhou, L.; Li, R.; Chen, L.; Shi, X. Investigation on low-temperature thermoelectric properties of Ag2Se polycrystal fabricated by using zone-melting method. *J. Phys. Chem. Lett.* **2021**, *12*, 8246–8255. [CrossRef]
26. Mao, J.; Zhu, H.; Ding, Z.; Liu, Z.; Gamage, G.A.; Chen, G.; Ren, Z. High thermoelectric cooling performance of n-type Mg_3Bi_2-based materials. *Science* **2019**, *365*, 495–498. [CrossRef]
27. Xu, Q.; Ming, C.; Xing, T.; Qiu, P.; Xiao, J.; Shi, X.; Chen, L. Thermoelectric properties of phosphorus-doped van der Waals crystal Ta4SiTe4. *Mater. Today Phys.* **2021**, *19*, 100417. [CrossRef]
28. Wang, L.; Li, J.; Zhang, C.; Ding, T.; Xie, Y.; Li, Y.; Lin, F.; Ao, W.; Zhang, C. Discovery of low-temperature GeTe-based thermoelectric alloys with high performance competing with Bi_2Te_3. *J. Mater. Chem. A* **2020**, *8*, 1660–1667. [CrossRef]
29. Narducci, D.; Giulio, F. Recent advances on thermoelectric silicon for low-temperature applications. *Materials* **2022**, *15*, 1214. [CrossRef]
30. Chung, D.-Y.; Hogan, T.P.; Rocci-Lane, M.; Brazis, P.; Ireland, J.R.; Kannewurf, C.R.; Bastea, M.; Uher, C.; Kanatzidis, M.G. A new thermoelectric material: $CsBi_4Te_6$. *J. Am. Ceram. Soc.* **2004**, *126*, 6414–6428.
31. Ramakrishnan, A.; Raman, S.; Chen, L.C.; Chen, K.-H. Enhancement in thermoelectric properties of TiS_2 by Sn addition. *J. Electron. Mater.* **2018**, *47*, 3091–3098. [CrossRef]
32. Whittingham, M.S. Lithium titanium disulfide cathodes. *Nature Energy* **2021**, *6*, 214. [CrossRef]
33. Wen, J.; Zhang, W.H.; Zhang, L.R.; Zhang, X.T.; Yu, Y.-X. Identification of the different contributions of pseudocapacitance and quantum capacitance and their electronic-structure-based intrinsic transport kinetics in electrode materials. *Chem. Phys. Lett.* **2021**, *775*, 138666. [CrossRef]
34. Ondes, B.; Evli, S.; Sahin, Y.; Uygun, M.; Uygun, D.A. Uricase based amperometric biosensor improved by AuNPs-TiS2 nanocomposites for uric acid determination. *Microchem. J.* **2022**, *181*, 107725. [CrossRef]
35. Gu, Y.; Song, K.; Hu, X.; Chen, C.; Pan, L.; Lu, C.; Shen, X.; Koumoto, K.; Wang, Y. Realization of an ultrahigh power factor and enhanced thermoelectric performance in TiS_2 via microstructural texture engineering. *ACS Appl. Mater. Interfaces* **2020**, *12*, 41687–41695. [CrossRef] [PubMed]
36. Gu, Y.; Song, K.; Hu, X.; Chen, C.; Pan, L.; Lu, C.; Shen, X.; Koumoto, K.; Wang, Y. Distinct anisotropy and a high-power factor in highly textured TiS_2 ceramics via mechanical exfoliation. *Chem. Commun.* **2020**, *56*, 5961–5964. [CrossRef]

37. Zhang, M.; Zhang, C.; You, Y.; Xie, H.; Chi, H.; Sun, Y.; Liu, W.; Su, X.; Yan, Y.; Tang, X.; et al. Electron density optimization and the anisotropic thermoelectric properties of Ti self-intercalated $Ti_{1+x}S_2$ compounds. *ACS Appl. Mater. Interfaces* **2018**, *10*, 32344–32354. [CrossRef]
38. Aoki, T.; Wan, C.; Ishiguro, H.; Morimitsu, H.; Koumoto, K. Evaluation of layered TiS_2-based thermoelectric elements fabricated by a centrifugal heating technique. *J. Ceram. Soc. Japan* **2011**, *119*, 382–385. [CrossRef]
39. Zhang, J.; Ye, Y.; Li, C.; Yang, J.; Zhao, H.; Xu, X.; Huang, R.; Pan, L.; Lu, C.; Wang, Y. Thermoelectric properties of TiS_2-$xPbSnS_3$ nanocomposites. *J. Alloys Compds.* **2017**, *696*, 1342–1348. [CrossRef]
40. Tian, R.; Wan, C.; Wang, Y.; Wei, Q.; Ishida, T.; Yamamoto, A.; Tsuruta, A.; Shin, W.; Li, S.; Koumoto, K. A solution-processed TiS_2/organic hybrid superlattice film towards flexible thermoelectric devices. *J. Mater. Chem. A* **2017**, *5*, 564–570. [CrossRef]
41. Wan, C.; Gu, X.; Dang, F.; Itoh, T.; Wang, Y.; Sasaki, H.; Kondo, M.; Koga, K.; Yabuki, K.; Snyder, G.J.; et al. Flexible n-type thermoelectric materials by organic intercalation of layered transition metal dichalcogenide TiS_2. *Nature Mater.* **2015**, *14*, 622–627. [CrossRef]
42. Ferhat, S.; Domain, C.; Vidal, J.; Noël, D.; Ratier, B.; Lucas, B. Flexible thermoelectric device based on $TiS_2(HA)_x$ n-type nanocomposite printed on paper. *Org. Electron.* **2019**, *28*, 256–263. [CrossRef]
43. Ursi, F.; Virga, S.; Garcia-Espejo, G.; Masciocchi, N.; Martorana, A.; Giannici, F. Long-term stability of TiS_2-alkylamine hybrid materials. *Materials* **2022**, *15*, 8297. [CrossRef]
44. Dua, Y.; Xu, J.; Paul, B.; Eklund, P. Flexible thermoelectric materials and devices. *Appl. Mater. Today* **2018**, *12*, 366–388. [CrossRef]
45. Jacob, S.; Delatouche, B.; Péré, D.; Khan, Z.U.; Ledoux, M.J.; Crispin, X.; Chmielowski, R. High-performance flexible thermoelectric modules based on high crystal quality printed TiS_2/hexylamine. *Sci. Technol. Adv. Mater.* **2021**, *22*, 907–916. [CrossRef]
46. Wan, C.; Kodama, Y.; Kondo, M.; Sasai, R.; Qian, X.; Gu, X.; Koga, K.; Yabuki, K.; Yang, R.; Koumoto, K. Dielectric Mismatch Mediates Carrier Mobility in Organic-Intercalated Layered TiS_2. *Nano Lett.* **2015**, *15*, 6302–6308. [CrossRef]
47. Govindaraj, P.; Sivasamy, M.; Murugan, K.; Venugopal, K.; Veluswamy, P. Pressure-driven thermoelectric properties of defect chalcopyrite structured $ZnGa_2Te_4$: Ab initio study. *RSC Adv.* **2022**, *12*, 12573–12582. [CrossRef]
48. Lee, H.S. *Thermoelectrics: Design and Materials*; John Wiley & Sons, Ltd.: New York, NY, USA, 2016.
49. Zhao, L.-D.; Lo, S.-H.; He, J.; Li, H.; Biswas, K.; Androulakis, J.; Wu, C.-I.; Hogan, T.P.; Chung, D.-Y.; Dravid, V.P.; et al. High performance thermoelectrics from earth-abundant materials: Enhanced figure of merit in PbS by second phase nanostructures. *J. Am. Chem. Soc.* **2011**, *133*, 20476–20487. [CrossRef]
50. Yin, S.; Wan, C.; Koumoto, K. High thermoelectric performance in flexible TiS_2/organic superlattices. *J. Ceram. Soc. Jpn.* **2022**, *130*, 211–218. [CrossRef]

Disclaimer/Publisher's Note: The statements, opinions and data contained in all publications are solely those of the individual author(s) and contributor(s) and not of MDPI and/or the editor(s). MDPI and/or the editor(s) disclaim responsibility for any injury to people or property resulting from any ideas, methods, instructions or products referred to in the content.

Communication

Multifunctional Cu-Se Alloy Core Fibers and Micro–Nano Tapers

Min Sun, Yu Liu, Dongdan Chen and Qi Qian *

State Key Laboratory of Luminescent Materials and Devices, Institute of Optical Communication Materials, Guangdong Provincial Key Laboratory of Fiber Laser Materials and Applied Techniques, and Guangdong Engineering Technology Research and Development Center of Special Optical Fiber Materials and Devices, School of Materials Science and Engineering, South China University of Technology, Guangzhou 510640, China
* Correspondence: qianqi@scut.edu.cn

Abstract: Cu-Se alloy core fibers with glass cladding were fabricated by a thermal drawing method of a reactive molten core. The composition, crystallography, and photoelectric/thermoelectric performance of the fiber cores were investigated. The X-ray diffraction spectra of the Cu-Se alloy core fibers illustrate the fiber cores being polycrystalline with CuSe and Cu_3Se_2 phases. Interestingly, the fiber cores show a lower electrical conductivity under laser irradiation than under darkness at room temperature. Meanwhile, the fiber cores possess a power factor of ~1.2 $mWm^{-1}K^{-2}$ at room temperature, which is approaching the value of the high thermoelectric performance bulk of Cu_2Se polycrystals. The flexible Cu-Se fibers and their micro–nano tapers have potential multifunctional applications in the field of photoelectric detection and thermoelectric conversion on curved surfaces.

Keywords: Cu-Se alloy; multifunctional fiber; thermal drawing; photoelectric detection; thermoelectric conversion

Citation: Sun, M.; Liu, Y.; Chen, D.; Qian, Q. Multifunctional Cu-Se Alloy Core Fibers and Micro–Nano Tapers. *Nanomaterials* **2023**, *13*, 773. https://doi.org/10.3390/nano13040773

Academic Editor: Jordi Sort

Received: 20 January 2023
Revised: 15 February 2023
Accepted: 17 February 2023
Published: 19 February 2023

Copyright: © 2023 by the authors. Licensee MDPI, Basel, Switzerland. This article is an open access article distributed under the terms and conditions of the Creative Commons Attribution (CC BY) license (https://creativecommons.org/licenses/by/4.0/).

1. Introduction

Copper selenide (CuSe) is a semiconductor material with excellent electrical and optical properties [1]. It is available in many crystalline phases and structures: the stoichiometric compounds of Cu_2Se, Cu_3Se_2, CuSe, or $CuSe_2$, and the non-stoichiometric $Cu_{2-x}Se$. Two-dimensional thin films of CuSe have been used in many applications, such as solar cells, photodetectors, thermoelectric devices, and gas sensors [2–7]. CuSe is reported to be hexagonal at room temperature, transforming to orthorhombic at 321 K, and returning to hexagonal at 393 K [8]. For $Cu_{1.8}Se$, the direct band gap width is 2.2 eV, and for Cu_2Se, the indirect band gap width is 1.4 eV [9]. However, the forbidden bandwidth of CuSe is 1.2 eV. There are various reasons for the wide variation of the band gap, including large stoichiometric deviations, a large number of dislocations, and quantum size effects [10]. In addition to studying the crystalline phases of CuSe, researchers have done many studies on the morphologies and electronic bands of CuSe compounds, including nanoparticles, nanotubes, and nanowires [11–15]. However, the reported CuSe materials only show sole-function applications and the multifunctional performance of CuSe materials should be systematically assessed.

In the field of photodetection, Kou et al. successfully prepared In^{3+}-doped $Cu_{2-x}Se$ nanostructures using an electrochemical deposition method [16]. It was analyzed that the doping of In^{3+} caused a larger photo-contact surface in the nanostructure, which facilitated the rapid separation of photogenerated charges, thus increasing the photocurrent. The results show that In^{3+} doping can improve the photoelectric properties of $Cu_{2-x}Se$ and has potential applications for photodetection devices. Furthermore, one of the extensive research fields of Cu-Se compounds is thermoelectric conversion. In 2013, Liu et al. investigated the phase transition properties of $Cu_{2-x}Se$, which led to improved thermal properties, and electrical properties by electron and phonon critical scattering [17]. In

$Cu_{2-x}Se$, Se forms a face-centered cubic sublattice, which provides a pathway for holes in the semiconductor. Cu ions are highly disordered around the Se sublattice and liquid-like in their mobility, resulting in $Cu_{2-x}Se$ having a very low thermal conductivity. It resulted in $Cu_{2-x}Se$ with ultra-high thermoelectric properties, and the highest ZT value of 2.3 at room temperature. $Cu_{2-x}Se$ has been expected to be used for microprocessor cooling and thermoelectric generators to power wireless sensors.

Herein, Cu-Se alloy core fibers were prepared with a thermal drawing method with a reactive molten core. The fiber cores showed a photoelectric response with a lower electrical conductivity under laser irradiation than under darkness. The fiber cores possessed high electrical conductivities and high Seebeck coefficients, approaching that of the bulk $Cu_{2-x}Se$ polycrystals [17]. The Cu-Se alloy core fibers with mechanical flexibility can be applied in flexible photoelectric detection, optical switching, thermoelectric sensing, and even multifunctional fiber sensing.

2. Experimental Procedure

The precursor powders of the fiber core were chosen to be CuSe. The CuSe used in this experiment was purchased from Longjin Materials Corporation, Shanghai, and it has a melting point of about 660 K and a density of about 6.8 g/cm^3. The CuSe powders were stacked into a BK7 glass tube purchased from Schott Corporation, Germany, and it possesses about a 1070 K softening temperature. Cu-Se core fibers with several meters in length were thermally drawn at about 1150 K in an optical fiber tower under argon. Additionally, some micro–nano Cu-Se fiber tapers were secondly drawn from the 200-μm-diameter Cu-Se fibers with an alcohol lamp.

The morphology and elemental distribution on the fiber end face were analyzed by an electron probe X-ray microanalysis (Shimadzu EPMA-1600). The crystallography of the cores was characterized by X-ray diffraction (XRD, X'Pert Pro) and microscopic Raman spectroscopy (Renishaw RM2000). The fiber cores were ground and analyzed by a UV-NIR spectrometer (Perkin-Elmer Lambda). The photocurrents of the fibers under 532/808 nm laser irradiation were measured by connecting both ends of the fiber to an external circuit [18]. Additionally, the Seebeck coefficient and electrical conductivity were measured by a two-probe method, when two ends of the fibers or tapers are silver-pasted and connected to the electrical circuit [19]. Under the same conditions, three-time measurements were performed to obtain the average measuring values, and their relative deviations were smaller than 5%.

3. Results and Discussion

Figure 1a shows the cross-section electron micrograph of the Gu-Se alloy core fiber. It can be seen that the core diameter is about 380 μm and the core/cladding structure is intact, indicating that the borosilicate glass and Cu-Se core have high-temperature wettability [20]. Figure 1b–e show the element distribution of the wavelength dispersive spectrometer (WDS) on the fiber end face. The boundary of the element distribution forms a circle, while Cu and Se are mainly distributed in the core region, and Si and O are mainly distributed in the cladding region. By the WDS of the electron probe micro analyzer (EPMA), the atomic ratio of Cu and Se was determined to be $Cu_{1.2}Se$. It is worth noting that both Figure 1b,c exhibit elemental enrichment, and the distribution of elements is not uniform. This implies that during the high-temperature drawing process, CuSe underwent some kind of Se element volatilization, and the atomic ratio of Cu and Se changed.

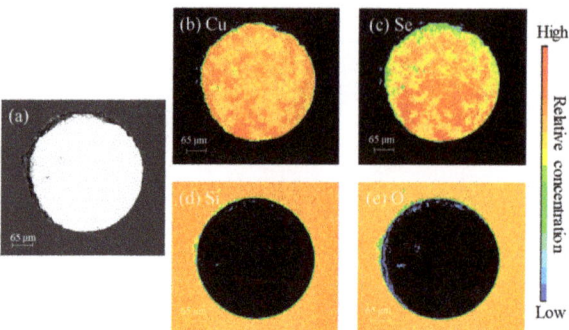

Figure 1. (a) Cross-section electronic micrograph of the Cu-Se core fiber; element distribution of (b) Cu, (c) Se, (d) Si, and (e) O in the fiber core region. The color palette at the right exhibits the relative concentration of the related element in (b–e) from low to high.

Figure 2a shows the XRD spectra of the fiber cores after drawing. It can be seen that the XRD peaks indicate the crystalline phases of Cu_3Se_2 (JCPDS#47-1745) and CuSe (JCPDS#34-0171). Figure 2b shows the Raman spectra of the precursor CuSe powders and Cu-Se fiber cores. In the range of 200–300 cm^{-1}, there is a peak at 262 cm^{-1} in the CuSe powders, which corresponds to the vibration of Cu-Se. The peak at 262 cm^{-1} is also present in the spectrum of the Cu-Se powders, and a new peak at 193.0 cm^{-1} is present. This peak position is consistent with that reported in the literature for Cu_3Se_2 [21], indicating that the Cu_3Se_2 crystalline phase was generated during the fiber drawing process, and that the final fiber crystalline phase composition was a mixture of CuSe and Cu_3Se_2.

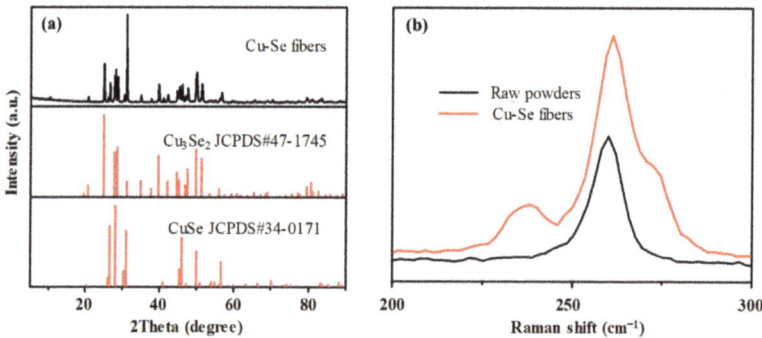

Figure 2. (a) X-ray diffraction spectra of the Gu-Se core fiber; and (b) Raman spectra of the Gu-Se raw powders and core.

The reflectance spectra of the CuSe core precursor powder and Cu-Se core composite glass fiber core can be obtained via a UV-NIR spectrometer, as shown in Figure 3. The CuSe raw powders and Cu-Se core powders have a strong absorption, and the strongest absorption is in the 500~1000 nm wavelength range. Therefore, visible light was chosen as the light source for the subsequent photocurrent test. The transmittance of Cu-Se core powders in this region is higher than that of the CuSe raw powders. The reason is that Cu_3Se_2 is generated during the CuSe fiber drawing process. The forbidden bandwidth is E_g = 2.03 eV for CuSe [22] and E_g = 1.45 eV for Cu_3Se_2 [23], due to the intrinsic absorption within the semiconductor material of sufficient energy for the photons to be absorbed. A photon of sufficient energy excites an electron, which leaps across the forbidden band into the empty conduction band and leaves a hole in the valence band. Due to the narrowing of the forbidden band width, photons can cross the forbidden band more easily, and the Cu-Se alloy core fiber can exhibit a stronger absorption of light. The photocurrents of the Cu-Se alloy core fiber in dark and irradiation conditions are carried out using the same experimental setup as our previous study [18], which was mainly focused on the thermoelectric and mechanical performance of Bi_2Te_3 micro–nano fibers without any optical irradiation. Figure 3b shows the current–voltage curves of the Cu-Se core fiber in dark and irradiation conditions. It can be seen that the electrical conductivity of the Cu-Se core fiber becomes significantly smaller under the irradiation of 532 nm laser and 808 nm laser. By linear fitting and calculation, the conductivity of the Cu-Se core fiber was reduced by 60% under the 532 nm laser and 80% under the 808 nm laser. According to the reported literature [22,23], the response of Cu_3Se_2 to light irradiation is mainly in the form of conductivity reduction.

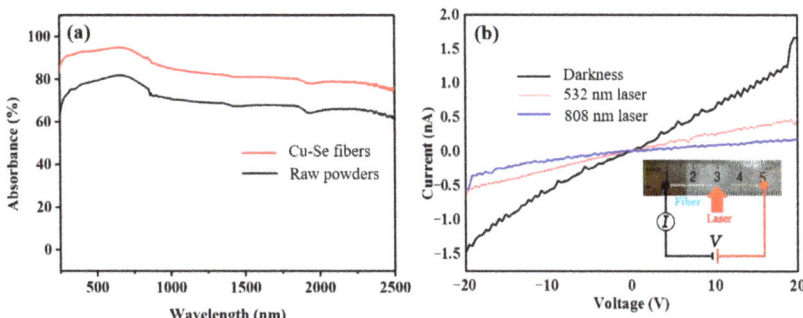

Figure 3. (a) Absorbance spectra for CuSe raw powders and Cu-Se core fibers; and (b) the current-voltage curves of a Cu-Se core fiber in dark and irradiated conditions. Inset of (b) is the optical graph of a Cu-Se fiber and schematic of the electrical circuit.

Beyond these, the Cu-Se alloy core fiber was secondly drawn by an alcohol lamp to be a micro–nano taper with a maximum diameter of about 200 μm and a minimum diameter lower than 1 μm. As shown in Figure 4a, two ends of a fiber or a taper were silver-pasted and connected to the electrical circuit. Additionally, when one end was heated at the groove and the other end was cooled at the sink, the thermoelectric voltage differences arose with temperature differences. In Figure 4b, the Seebeck coefficients of the Cu-Se fiber and its micro–nano taper were measured, being 101 μV/K and 117 μV/K, respectively, at room temperature. When the Cu-Se taper possesses a 16% higher Seebeck coefficient than the Cu-Se fiber does, it should be derived from a size-related effect. So the power factors ($PF = S^2\sigma$) of the fiber and taper were calculated, respectively, being about 1.2 mWm^{-1}K^{-2} and 1.6 mWm^{-1}K^{-2}, approaching that of the bulk Cu_2Se polycrystals, when the electrical conductivity of the taper was estimated to be 1176 S/cm, the same as the Cu-Se fiber cores. Additionally, the ZT values of the fibers and tapers were calculated, respectively, being approximately 0.45 and 0.60 at 300 K, when their thermal conductivity was estimated to be about 0.8 Wm^{-1}K^{-1}, as is reported for the bulk Cu_2Se polycrystals [17]. The ZT value of the fibers at 300 K was similar to that of the bulk Cu_2Se polycrystals [17], but much lower than the ZT value of 2.3 at their phase change point of 400 K. In addition, for estimating their mechanical flexibility [18], the Cu-Se fibers with a diameter (D) of 200 μm exhibited a minimum bending radius (r_{min}) of approximately 2 cm during the bending tests, and a maximum bending strain ($\varepsilon_{max} = D/2r_{min}$) of 0.5%. Therefore, Cu-Se alloy core fibers can be expected to be used as multifunctional fibers in the fields of photoelectric detection and thermoelectric conversion on curved surfaces.

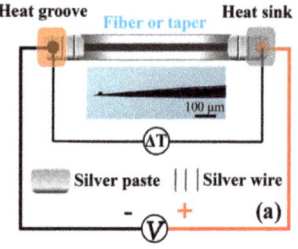 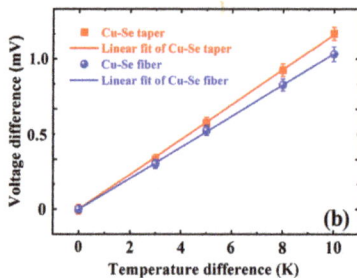

Figure 4. (**a**) Schematic of an electrical circuit connecting the two ends of a fiber or a micro–nano taper; the inset of (**a**) is the optical micrograph of a Cu-Se taper. (**b**) The thermoelectric voltages at temperature differences of 0 K, 3 K, 5 K, 8 K, or 10 K cross the two ends. The error bar in (**b**) shows 5% measurement uncertainty.

4. Conclusions

In conclusion, Cu-Se alloy core fibers have been fabricated by using molten-core thermal drawing. The polycrystalline Cu-Se cores are constituted of CuSe and Cu_3Se_2, and their composition is made up of $Cu_{1.2}Se$. Interestingly, the electrical conductivity of the Cu-Se cores under laser irradiation is only a third of that under darkness. The as-drawn Cu-Se core fibers possess a *PF* of ~1.2 mWm^{-1}K^{-2} and *ZT* of ~0.45. The secondly drawn Cu-Se core tapers possess a *PF* of ~1.6 mWm^{-1}K^{-2} and *ZT* of ~0.60. Ultimately, the Cu-Se core shows a good flexibility and photo-/thermo-electric responses at room temperature, and future work should focus on the related size effect, phase change, and optimized applications of these multifunctional fibers.

Author Contributions: Conceptualization, M.S., Q.Q.; methodology, M.S., Y.L.; writing—original draft preparation, M.S.; supervision, D.C., Q.Q. All authors have read and agreed to the published version of the manuscript.

Funding: The work is supported by the Natural Science Foundation of China (52002131), Key R&D Program of Guangzhou (202007020003).

Data Availability Statement: The production data are available on request from the corresponding author.

Conflicts of Interest: The authors declare no conflict of interest.

References

1. Liu, H.; Shi, X.; Xu, F.; Zhang, L.; Zhang, W.; Chen, L.; Li, Q.; Uher, C.; Day, T.; Snyder, G.J. Copper ion liquid-like thermoelectrics. *Nat. Mater.* **2012**, *11*, 422. [CrossRef]
2. Chen, W.S.; Stewart, J.M.; Mickelsen, R.A. Polycrystalline thin-film $Cu_{2-x}Se$/CdS solar cell. *Appl. Phys. Lett.* **1985**, *46*, 1095–1097. [CrossRef]
3. Wang, Z.; Peng, F.; Wu, Y.; Yang, L.; Zhang, F.; Huang, J. Template synthesis of $Cu_{2-x}Se$ nanoboxes and their gas sensing properties. *CrystEngComm* **2012**, *14*, 3528–3533. [CrossRef]
4. Korzhuev, M.A. Dufour effect in superionic copper selenide. *Phys. Solid State* **1998**, *40*, 217–219. [CrossRef]
5. Xu, J.; Zhang, W.; Yang, Z.; Ding, S.; Zeng, C.; Chen, L.; Wang, Q.; Yang, S. Large-Scale synthesis of long crystalline $Cu_{2-x}Se$ nanowire bundles by water-evaporation-induced self-assembly and their application in gas sensing. *Adv. Funct. Mater.* **2009**, *19*, 1759–1766. [CrossRef]
6. Jin, C.; Shevchenko, N.A.; Li, Z.; Popov, S.; Chen, Y.; Xu, T. Nonlinear coherent optical systems in the presence of equalization enhanced phase noise. *J. Light. Technol.* **2021**, *39*, 4646–4653. [CrossRef]
7. Wang, L.; Chen, J.; Liu, C.; Wei, M.; Xu, X. CuO-modified $PtSe_2$ monolayer as a promising sensing candidate toward C_2H_2 and C_2H_4 in oil-immersed transformers: A density functional theory study. *ACS Omega* **2022**, *7*, 45590–45597. [CrossRef]
8. Stevels, A.L.N.; Jellinek, F. Phase transitions in copper chalcogenides: I. The copper-selenium system. *Recl. Des Trav. Chim. Des Pays-Bas* **1971**, *90*, 273–283. [CrossRef]
9. Hermann, A.M.; Fabick, L. Research on polycrystalline thin-film photovoltaic devices. *J. Cryst. Growth* **1983**, *61*, 658–664. [CrossRef]

10. Saitoh, T.; Matsubara, S.; Minagawa, S. Polycrystalline indium phosphide solar cells fabricated on molybdenum substrates. *Jpn. J. Appl. Phys.* **1977**, *16*, 807. [CrossRef]
11. Liu, X.; Wang, X.; Zhou, B.; Law, W.C.; Cartwright, A.N.; Swihart, M.T. Size-controlled synthesis of $Cu_{2-x}E$ (E= S, Se) nanocrystals with strong tunable near-infrared localized surface plasmon resonance and high conductivity in thin films. *Adv. Funct. Mater.* **2013**, *23*, 1256–1264. [CrossRef]
12. Riha, S.C.; Johnson, D.C.; Prieto, A.L. Cu_2Se nanoparticles with tunable electronic properties due to a controlled solid-state phase transition driven by copper oxidation and cationic conduction. *J. Am. Chem. Soc.* **2010**, *133*, 1383–1390. [CrossRef]
13. Dorfs, D.; Hartling, T.; Miszta, K.; Bigall, N.C.; Kim, M.R.; Genovese, A.; Falqui, A.; Povia, M.; Manna, L. Reversible tunability of the near-infrared valence band plasmon resonance in $Cu_{2-x}Se$ nanocrystals. *J. Am. Chem. Soc.* **2011**, *133*, 11175–11180. [CrossRef]
14. Zhang, S.Y.; Fang, C.X.; Tian, Y.P.; Zhu, K.R.; Jin, B.K.; Shen, Y.H.; Yang, J.X. Synthesis and characterization of hexagonal CuSe nanotubes by templating against trigonal Se nanotubes. *Cryst. Growth Des.* **2006**, *6*, 2809–2813. [CrossRef]
15. Xu, J.; Tang, Y.B.; Chen, X.; Luan, C.-Y.; Zhang, W.-F.; Zapien, J.A.; Zhang, W.-J.; Kwong, H.-L.; Meng, X.-M.; Lee, S.-T.; et al. Synthesis of homogeneously alloyed $Cu_{2-x}(S_ySe_{1-y})$ nanowire bundles with tunable compositions and bandgaps. *Adv. Funct. Mater.* **2010**, *20*, 4190–4195. [CrossRef]
16. Kou, H.; Jiang, Y.; Li, J.; Yu, S.; Wang, C. Enhanced photoelectric performance of $Cu_{2-x}Se$ nanostructure by doping with In^{3+}. *J. Mater. Chem.* **2012**, *22*, 1950–1956. [CrossRef]
17. Liu, H.; Yuan, X.; Lu, P.; Shi, X.; Xu, F.; He, Y.; Tang, Y.; Bai, S.; Zhang, W.; Chen, L.; et al. Ultrahigh Thermoelectric performance by electron and phonon critical scattering in $Cu_2Se_{1-x}I_x$. *Adv. Mater.* **2013**, *25*, 6607–6612. [CrossRef]
18. Sun, M.; Tang, G.; Wang, H.; Zhang, T.; Zhang, P.; Han, B.; Yang, M.; Zhang, H.; Chen, Y.; Chen, J.; et al. enhanced thermoelectric properties of Bi_2Te_3-based micro–nano fibers via thermal drawing and interfacial engineering. *Adv. Mater.* **2022**, *34*, 2202942. [CrossRef]
19. Sun, M.; Tang, G.; Huang, B.; Chen, Z.; Zhao, Y.-J.; Wang, H.; Zhao, Z.; Chen, D.; Qian, Q.; Yang, Z. Tailoring microstructure and electrical transportation through tensile stress in Bi_2Te_3 thermoelectric fibers. *J. Mater.* **2020**, *6*, 467.
20. Sun, M.; Qian, Q.; Tang, G.; Liu, W.; Qian, G.; Shi, Z.; Huang, K.; Chen, D.; Xu, S.; Yang, Z. Enhanced thermoelectric properties of polycrystalline Bi_2Te_3 core fibers with preferentially oriented nanosheets. *APL Mater.* **2018**, *6*, 036103. [CrossRef]
21. Monjezi, F.; Jamali-Sheini, F.; Yousefi, R. Pb-doped Cu_3Se_2 nanosheets: Electrochemical synthesis, structural features and optoelectronic properties. *Sol. Energy* **2018**, *171*, 508–518. [CrossRef]
22. Gosavi, S.R.; Deshpande, N.G.; Gudage, Y.G.; Sharma, R. Physical, optical and electrical properties of copper selenide (CuSe) thin films deposited by solution growth technique at room temperature. *J. Alloy. Compd.* **2008**, *448*, 344–348. [CrossRef]
23. Qiao, L.N.; Wang, H.C.; Shen, Y.; Lin, Y.H.; Nan, C.W. Enhanced photocatalytic performance under visible and near-infrared irradiation of $Cu_{1.8}Se/Cu_3Se_2$ composite via a phase junction. *Nanomaterials* **2017**, *7*, 19. [CrossRef]

Disclaimer/Publisher's Note: The statements, opinions and data contained in all publications are solely those of the individual author(s) and contributor(s) and not of MDPI and/or the editor(s). MDPI and/or the editor(s) disclaim responsibility for any injury to people or property resulting from any ideas, methods, instructions or products referred to in the content.

Communication

High-Performance n-Type Bi$_2$Te$_3$ Thermoelectric Fibers with Oriented Crystal Nanosheets

Min Sun [1], Pengyu Zhang [1,2], Guowu Tang [1,3], Dongdan Chen [1,*], Qi Qian [1,*] and Zhongmin Yang [1]

[1] State Key Laboratory of Luminescent Materials and Devices, Institute of Optical Communication Materials, Guangdong Provincial Key Laboratory of Fiber Laser Materials and Applied Techniques, and Guangdong Engineering Technology Research and Development Center of Special Optical Fiber Materials and Devices, School of Materials Science and Engineering, South China University of Technology, Guangzhou 510640, China

[2] Nanjing Institute of Future Energy System, Nanjing 211135, China

[3] School of Physics and Optoelectronic Engineering, Guangdong University of Technology, Guangzhou 510006, China

* Correspondence: ddchen@scut.edu.cn (D.C.); qianqi@scut.edu.cn (Q.Q.)

Abstract: High-performance thermoelectric fibers with n-type bismuth telluride (Bi$_2$Te$_3$) core were prepared by thermal drawing. The nanosheet microstructures of the Bi$_2$Te$_3$ core were tailored by the whole annealing and Bridgman annealing processes, respectively. The influence of the annealing processes on the microstructure and thermoelectric performance was investigated. As a result of the enhanced crystalline orientation of Bi$_2$Te$_3$ core caused by the above two kinds of annealing processes, both the electrical conductivity and thermal conductivity could be improved. Hence, the thermoelectric performance was enhanced, that is, the optimized dimensionless figure of merit (ZT) after the Bridgman annealing processes increased from 0.48 to about 1 at room temperature.

Keywords: n-type Bi$_2$Te$_3$; thermoelectric fibers; thermal drawing; crystalline orientation

---

Citation: Sun, M.; Zhang, P.; Tang, G.; Chen, D.; Qian, Q.; Yang, Z. High-Performance n-Type Bi$_2$Te$_3$ Thermoelectric Fibers with Oriented Crystal Nanosheets. *Nanomaterials* **2023**, *13*, 326. https://doi.org/10.3390/nano13020326

Academic Editor: Seung Hwan Ko

Received: 14 December 2022
Revised: 6 January 2023
Accepted: 10 January 2023
Published: 12 January 2023

Copyright: © 2023 by the authors. Licensee MDPI, Basel, Switzerland. This article is an open access article distributed under the terms and conditions of the Creative Commons Attribution (CC BY) license (https://creativecommons.org/licenses/by/4.0/).

1. Introduction

Thermoelectrics (TEs) can directly convert heat to electricity through the directional movement of the internal carriers with the temperature difference. Due to their small size, high reliability, and no noise, thermoelectric devices have great potential for civil and military applications [1,2]. The dimensionless figure of merit (ZT) is used to identify the performance of TEs, which is determined by the inherent electrical conductivities (σ), Seebeck coefficients (S), power factors (PF = $S^2\sigma$), and thermal conductivities (κ). Nowadays, bismuth telluride (Bi$_2$Te$_3$) is considered a class of state-of-the-art TEs (ZT > 1) for low-temperature applications (0–300 °C), such as Peltier refrigerators or CCD coolers [3]. Due to its anisotropic structure, Bi$_2$Te$_3$ is considered as a significant candidate for directional thermoelectric properties, such as the *c* plane or across the out-of-plane of Bi$_2$Te$_3$ sheets [4,5]. Theoretically, the ability of crystalline orientation is highly related to the crystal size and orientation degree of the *c* plane (F) [6]. There have been many studies on the introduction of crystalline orientation at various scales [7–12]; however, the artificial regulation of crystalline orientation and related mechanisms have rarely been reported.

In high-performance thermoelectric fiber devices, p-n Bi$_2$Te$_3$ fibers are usually paired in series electrically and in parallel thermally to realize thermoelectric conversion [13]. To date, it has been found that the n-type Bi$_2$Te$_3$ fibers (thermally drawn with glass cladding by a powder-in-tube method) could exhibit a restricted ZT < 0.5 owing to the low relative density and more texturing-related sensitivity of the carriers' mobility than their p-type fibers (ZT~1.4) [9]. In the thermoelectric fibers fabricated by thermal drawing from our previous works [14–17], the *c*-plane crystalline orientation enhances the carrier mobility and electrical conductivity in Bi$_2$Te$_3$ core fibers.

Herein, high-performance n-type Bi_2Te_3 fibers were fabricated by a rod-in-tube thermal drawing method, and subsequent annealing processes (whole annealing and Bridgman annealing) were used to enhance the crystallization of two kinds of opposite crystalline orientations [17]. Crystalline orientations along (110) and (001) in the fiber cores were increased by whole annealing and Bridgman annealing, respectively, and the elemental enrichment was reduced. Hence, the resulting fibers exhibit tailored electrical-phonon transport, and the Bridgman-annealed fibers show a better thermoelectric performance. The optimized ZT values of the annealed fibers are measured to be about 1 near room temperature, which is about twice as large as that of their as-drawn fiber counterparts. The n-type Bi_2Te_3 fibers with oriented crystal nanosheets in the core and the glass cladding protection can be connected by electroconductive paste electrically in series and thermally in parallel with the similar p-type Bi_2Te_3 fibers to fabricate thermoelectric elements [6,9,13]. Additionally, they could be used in the field of waste heat recycling on the curved surface (e.g., hot water tubes and vehicle tailpipes), Peltier cooling, and temperature-sensing textiles (e.g., face masks and sleeveless shirts), etc.

2. Experimental Procedure

2.1. Fabrication

A two-step method with thermal drawing and post-processing annealing was applied on n-type Bi_2Te_3 fibers fabrication. The Bi_2Te_3 and Bi_2Se_3 powders of 99.999% purity (under 200 mesh, Aladdin, Ontario, CA, USA) at a ratio of 9:1 were used as raw materials to prepare an n-type Bi_2Te_3 rod with a relative density of 99%, and the fabrication process is reported in detail in our previous study [16]. The Bi_2Te_3 rod was inserted into a borosilicate glass (BK7, Schott, Zagreb, Croatia) tube, which has a glass-transition temperature of 562 °C and a softening temperature of 800 °C, forming a fiber preform. Several-meter-length Bi_2Te_3-core/glass-clad fibers were drawn from the fiber preform at ~880 °C in an optical fiber drawing tower under an argon atmosphere.

One group of the as-drawn fibers were wholly annealed at 565 °C for 5 h in a muffle furnace and cooled down to room temperature at a rate of 0.1 °C/min. The other group of the as-drawn fibers were Bridgman-method annealed, which descended step by step and crossed a high-temperature ring with a constant speed of 1 cm/h to recrystallize the core as shown in our previous study [6]. The ring-zone temperature is ~645 °C, which is higher than the melting temperature of the Bi_2Te_3 ~585 °C.

2.2. Measurements

For characterizing the crystallinity and electrical transport of the fiber cores, the as-drawn and annealed fibers were etched in HF acid to strip the glass cladding, and then, the fiber cores were identified by X-ray diffractometer (XRD, X'Pert PROX, Cu Kα). Energy-dispersive spectroscopy (EDS) of elemental studies was carried out on the fiber cross-sections by using scanning electron microscopy (SEM, Zeiss Merlin, Oberkochen, Germany).

The Seebeck coefficients (S) and the electrical conductivities (σ) of the fiber cores were tested by a four-probe method [9]. Three-time measurements were performed under the same conditions for each fiber core to obtain the average testing value, and the relative deviations were <5%. The thermal conductivity (κ) was tested by a method of time-domain thermal reflection, and the relative deviation was <10%. All the relative deviations show the testing results are reproducible and reliable.

3. Results and Discussion

3.1. Microstructure

The XRD patterns of as-drawn fiber cores, whole-annealed fiber cores, and Bridgman-annealed fiber cores are exhibited in Figure 1. All the XRD peaks are indexed to the $Bi_2Te_{2.7}Se_{0.3}$ hexagonal phase (JCPDS#50-0954). The as-drawn $Bi_2Te_{2.7}Se_{0.3}$ fiber cores (BTSF) are polycrystals after a process of thermal drawing and quick cooling (>100 °C/s). The average crystal size of as-drawn fiber core is estimated to be ~30 nm based on the XRD peak width and the Scherrer formula [16], and the average particle sizes of post-annealed fiber cores are >100 nm. A great difference is found in XRD peaks among the as-drawn fibers and annealed fibers. The annealed fiber cores show larger XRD peak intensities than the as-drawn fiber cores at special lattice planes, such as (1 1 0) marked in blue or (0 0 6) marked in red. The difference in XRD peak intensities demonstrates that the crystals in annealed fiber cores could possess an orientation over the crystals in as-drawn fiber cores. According to the Lotgering method [6], the orientation degree F of the (0 0 1) planes in polycrystals can be calculated: $F = (P - P_0) / (1 - P_0)$; $P_0 = I_0(0\,0\,l) / \sum I_0(h\,k\,l)$; $P = I(0\,0\,l) / \sum I(h\,k\,l)$. Hence, the F of the crystals in as-drawn fiber cores and annealed fiber cores are calculated as 0.45, 0.06, and 0.92, respectively. This means that the as-drawn fiber cores and Bridgman-annealed fiber cores are in the (0 0 1) orientation and that the Bridgman-annealed fiber cores show a greater orientation. Oppositely, the whole-annealed fiber cores show a great (1 1 0) orientation when their (110) orientation factor ($F_{(110)}$~0.85) can be calculated from the data of Figure 1.

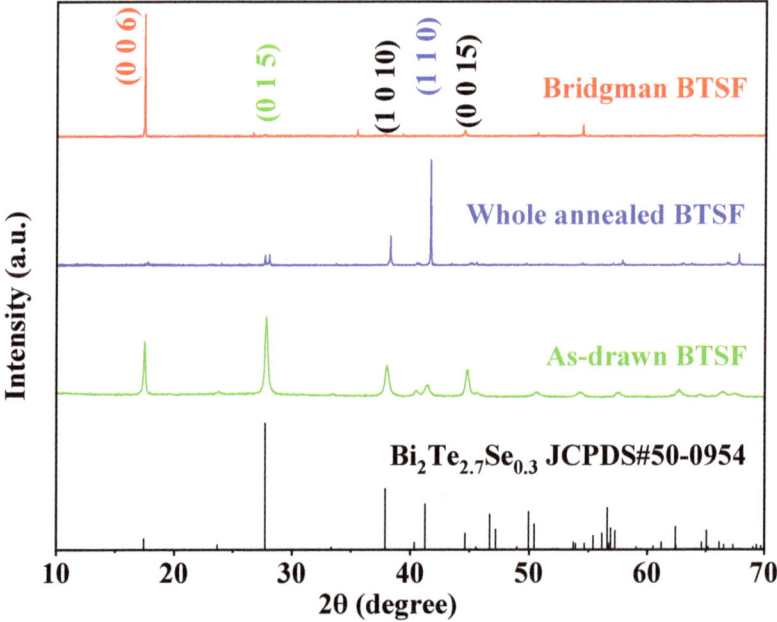

Figure 1. XRD patterns of the as-drawn and annealed BTSF.

The cross-section SEM images and EDS elemental mappings of Bi, Se, and Te on the as-drawn and annealed fibers are shown in Figure 2. All three samples show a layered structure in the cross-section. In Figure 2a,i, it is shown that the three samples show similar nanosheet microstructures, following the (001) orientation of the as-drawn/Bridgman-annealed fiber cores. It can be carefully observed that the nanosheets are almost parallel to the cross-section at the bottom left of the whole-annealed fiber core in Figure 2e, following the (110) orientation of the whole-annealed fiber cores. In the elemental mappings, there are traces of elements Bi, Se, and Te diffused from the core into the cladding region. It is observed that there are Te enrichments in the as-drawn fiber core, but little enrichment can be found in the annealed fiber cores. Enriched Te could be decreased by whole annealing at 565 °C or Bridgman annealing at 645 °C with a 1 cm/h recrystallization speed. Since the crystalline orientation and Te enrichments impact the electrical-phonon transport [18], the fiber with different Te enrichments may show diverse TE performance.

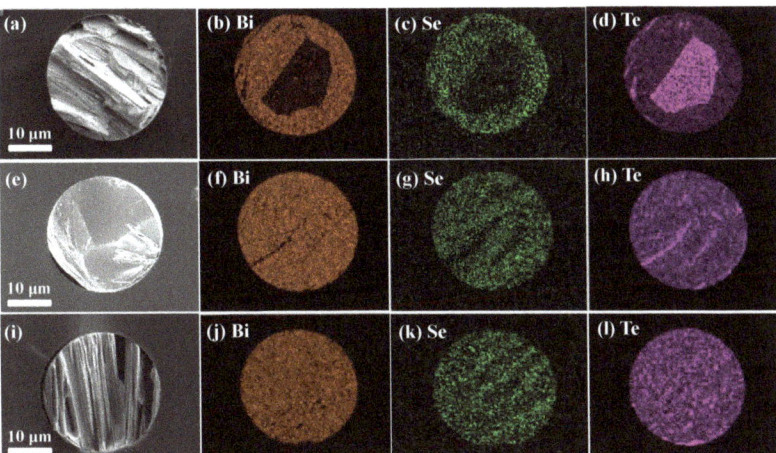

Figure 2. SEM cross-sectional images and EDS elemental mappings of (**a**–**d**) the as-drawn, (**e**–**h**) the whole-annealed, and (**i**–**l**) the Bridgman-annealed fibers.

In the micrometer-scale limited space of fiber cores, it is important to uncover the growth process and mechanism of the oppositely crystalline orientation, whose nanosheet microstructure and element distribution are shown in Figure 2. As the as-drawn fibers went through a quick cooling process (>100 °C/s) after thermal drawing, as reported in our previous work [6], the fiber cores possessed a preferred (001) orientation because of residual thermal stress along the radial direction from the core to the cladding. The microstructure evolutions of the as-drawn fibers during annealing are schematically illustrated in Figure 3. In the whole annealing process (i), the thermal stress gradually decreased by annealing and slow cooling. The fiber core then exhibits a trend of being a preferred (110) orientation with the thermal stress decreasing. In the Bridgman annealing process (ii), the core melt gradually recrystallizes into nanosheets around the fiber interface to minimize surface energy. In the limited fiber core space, the nanosheets recrystallize continuously along the (001) plane and the fiber axis. It should be illuminated in a similar microstructural orientation during a single-crystal crystallization process by directional Bridgman annealing [19] or laser annealing [20]. In addition, there is elemental diffusion on the core–clad interface, which might cause surface roughness and low-dimensional defects of the fiber core, as in our previous study [9].

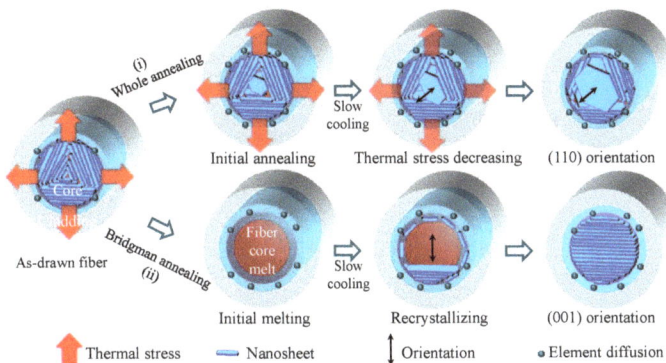

Figure 3. Two oriented crystal growth models of the n-type Bi_2Te_3 fiber core during whole annealing or Bridgman annealing.

3.2. Thermoelectric Properties

The electrical conductivities (σ) and Seebeck coefficients (S) of the fibers are shown in Figure 4a,b, measured by a home-made setup in our previous work [6]. In Figure 4a, all the as-drawn/annealed fibers possess a decreasing σ with increasing temperature (10–50 °C), revealing the metallic resistance characteristic [21]. The σ_w of the whole-annealed fibers and the σ_b of the Bridgman-annealed fibers are higher than the σ_a of the as-drawn fibers, and the σ_b is approaching threefold σ_a at the same temperature. The result could be derived from the fact that the (001) orientation of Bridgman-annealed fibers or the (110) orientation of whole-annealed fibers could enhance σ, and the greater orientation along (001) would support the higher carrier mobility and σ, as reported in our previous work [6]. In Figure 4b, the negative Seebeck coefficient means that all the fibers are n-type conductors [22]. All the as-drawn fibers and the annealed fibers possess an increasing $|S|$ with increasing temperature (10–50 °C). The $|S_w|$ of the whole-annealed fibers is bigger than $|S_b|$ of the Bridgman-annealed fibers and $|S_a|$ of the as-drawn fibers under the same temperature. This could be attributed to the fact that the annealing process could decrease Te enrichment to increase $|S|$, and the oriented crystals along (110) should exhibit a little higher $|S|$ than the oriented crystals along (001), whose isotropic behavior is following the result of the reported n-type Bi_2T_3 single crystals [23] or films [24]. The $PF = S^2\sigma$ of all fibers were calculated and shown in Figure 4c. The PF_b of the Bridgman-annealed fibers is higher than the PF_w of the whole-annealed fibers and the PF_a of the as-drawn fibers at the same temperature. The PF_b of the Bridgman-annealed fibers exceeds fourfold the PF_a, and the highest value is about 4 mW/mK^2 at 10 °C.

The measured σ, S, κ, and calculated ZT of all three samples are listed in Table 1. For the thermal conductivities, the κ_a of the as-drawn fiber is ultralow, which could be caused by enhanced phonon scattering from the nanocrystalline grains, as reported in our previous work [16]. The room temperature ZT (~27 °C) of the Bridgman-annealed $Bi_2Te_{2.7}Se_{0.3}$ fibers is the highest, which is twice as large as that of the as-drawn fibers and is two times larger than the reported ZT of the $Bi_2Te_{2.5}Se_{0.5}$ or Bi_2Se_3 fibers [13,15]. Even the Bridgman-annealed fibers possess a higher ZT than that of the Bridgman-growth-method Bi-Te-Se crystals [25], which shows higher σ and κ but a lower S at the end of Table 1. For the high ZT, the Bridgman-annealed fibers possess a higher σ and a higher κ than $Bi_2Te_{2.5}Se_{0.5}$ fibers, which benefits from the (001) orientation. Additionally, the whole-annealed fibers possess a higher σ but a similar κ, which also benefits from the (110) orientation. Different from the reported works found in oriented Bi_2Te_3 single crystals [3,23], whose ZT doubles in the (001) compared to along the (110), the ZT of n-type Bi_2Te_3 fibers is similar in the (001) orientation by Bridgman annealing compared to along the (110) orientation by whole annealing. In the meantime, during the bending test for estimating mechanical flexibility [9], the annealed fibers with a diameter D of 200 μm exhibit a minimum bending radius (r_{min}) of about 1.8 cm and a maximum bending strain ($\varepsilon_{max} = D / 2r_{min}$) of 0.56%.

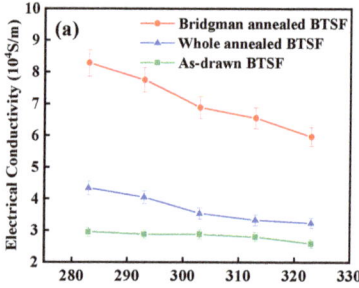

Figure 4. Cont.

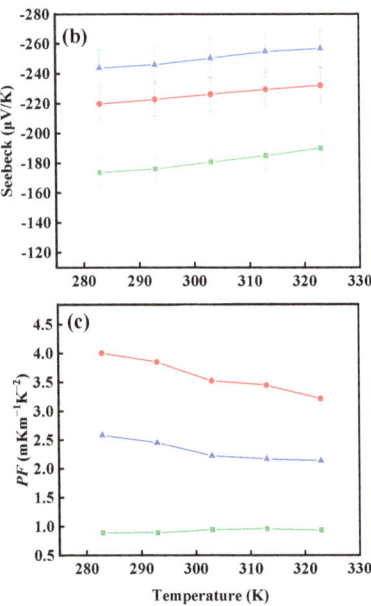

Figure 4. (a) The σ, (b) S, and (c) calculated PF of the as-drawn fibers and the annealed fibers at 10–50 °C.

Table 1. σ, S, κes, and calculated ZT of the n-type Bi_2Te_3-based fibers and Bridgman-method-growth crystals near room temperature.

Samples	Electrical Conductivity σ (S/cm)	Seebeck Coefficient S (μV/K)	Thermal Conductivity κ (W/mK)	ZT
Bridgman-annealed fibers	689 ± 33	−226 ± 11	1.01 ± 0.1	1.05
Whole-annealed fibers	355 ± 17	−251 ± 12	0.71 ± 0.07	0.95
As-drawn fibers	289 ± 14	−181 ± 9	0.59 ± 0.06	0.48
$Bi_2Te_{2.5}Se_{0.5}$ core fibers [17]	180	−227	0.64	0.43
Bi_2Se_3 fibers [13]	763	−92	0.84	0.23
Bi_2Se_3 core fibers [15]	319	−150	1.25	0.18
Bridgman Bi-Te-Se crystals [25]	1064	−201	1.4	0.92

4. Conclusions

In conclusion, high TE performance n-type Bi_2Te_3 fibers were prepared via rod-in-tube thermal drawing and Bridgman/whole annealing. The polycrystalline Bi_2Te_3 cores possess a polycrystalline orientation, and the F of (001) orientation factor in the Bridgman-annealed fiber cores is 0.92, and the F of (110) orientation factor in the whole-annealed fiber cores is 0.85. The (001) orientation increases the electrical conductivity and the thermal conductivity of the fiber cores more than the (110) orientation. Interestingly, both ZT of n-type Bi_2Te_3 fibers along (001) orientation and (110) orientation are approximately 1. Additionally, our future work will apply X-ray microfluorescence microscopy or tomography to study the composition of fibers in length and diameter. Finally, the Bridgman-annealed Bi_2Te_3 core shows an enhanced $ZT = 1.05$ at room temperature, and our future work will be on the TE device applications of these fibers. This proof-of-concept method of thermal drawing and annealing has potential in fiber-based TE applications.

Author Contributions: Conceptualization, M.S., D.C., Q.Q.; methodology, M.S., P.Z.; writing—original draft preparation, M.S.; writing—review and editing, G.T., D.C., Q.Q.; supervision, D.C., Q.Q., Z.Y. All authors have read and agreed to the published version of the manuscript.

Funding: This research was funded by Natural Science Foundation of China (52002131, 62005080, 52172249), 2021 Talent Revitalization Plan Project for New High Performance Material Industry in Qingyuan City (2021YFJH02001), Local Innovative and Research Teams Project of Guangdong Pearl River Talents Program (2017BT01X137), Key R&D Program of Guangzhou (202007020003), and Carbon Peaking Carbon Neutrality Science and Technology Innovation of Jiangsu Province (BE2022011).

Data Availability Statement: The production data are available on request from the corresponding author.

Conflicts of Interest: The authors declare no conflict of interest.

References

1. Tarancón, A. Powering the IoT revolution with heat. *Nat. Electron.* **2019**, *2*, 270–271. [CrossRef]
2. Wu, Z.; Hu, Z. Perspective—Powerful Micro/Nano-Scale Heat Engine: Thermoelectric Converter on Chip. *ECS Sens. Plus* **2022**, *1*, 023402. [CrossRef]
3. Yox, P.; Viswanathan, G.; Sarkar, A.; Wang, J.; Kovnir, K. Thermoelectric Materials. In *Reference Module in Chemistry, Molecular Sciences and Chemical Engineering*; Elsevier: Amsterdam, The Netherlands, 2022; Available online: https://linkinghub.elsevier.com/retrieve/pii/B9780128231449001096 (accessed on 13 December 2022).
4. Pei, J.; Cai, B.; Zhuang, H.L.; Li, J.-F. Bi_2Te_3-based applied thermoelectric materials: Research advances and new challenges. *Natl. Sci. Rev.* **2020**, *7*, 1856–1858. [CrossRef] [PubMed]
5. Shen, Y.; Wang, C.; Yang, X.; Li, J.; Lu, R.; Li, R.; Zhang, L.; Chen, H.; Zheng, X.; Zhang, T. New Progress on Fiber-Based Thermoelectric Materials: Performance, Device Structures and Applications. *Materials* **2021**, *14*, 6306. [CrossRef] [PubMed]
6. Sun, M.; Tang, G.; Huang, B.; Chen, Z.; Zhao, Y.-J.; Wang, H.; Zhao, Z.; Chen, D.; Qian, Q.; Yang, Z. Tailoring microstructure and electrical transportation through tensile stress in Bi_2Te_3 thermoelectric fibers. *J. Mater.* **2020**, *6*, 467–475. [CrossRef]
7. Zhu, B.; Liu, X.; Wang, Q.; Qiu, Y.; Shu, Z.; Guo, Z.; Tong, Y.; Cui, J.; Guc, M.; He, J. Realizing record high performance in n-type Bi_2Te_3-based thermoelectric materials. *Energy Environ. Sci.* **2020**, *13*, 2106–2114. [CrossRef]
8. Zheng, Z.-H.; Shi, X.-L.; Ao, D.-W.; Liu, W.-D.; Li, M.; Kou, L.-Z.; Chen, Y.-X.; Li, F.; Wei, M.; Liang, G.-X.; et al. Harvesting waste heat with flexible Bi_2Te_3 thermoelectric thin film. *Nat. Sustain.* **2022**, *1501*, 1–12. [CrossRef]
9. Sun, M.; Tang, G.; Wang, H.; Zhang, T.; Zhang, P.; Han, B.; Yang, M.; Zhang, H.; Chen, Y.; Chen, J.; et al. Enhanced thermoelectric properties of Bi_2Te_3-based micro-nano fibers via thermal drawing and interfacial engineering. *Adv. Mater.* **2022**, *34*, 2202942. [CrossRef]
10. Zhang, J.; Zhang, H.; Wang, Z.; Li, C.; Wang, Z.; Li, K.; Huang, X.; Chen, M.; Chen, Z.; Tian, Z.; et al. Single-crystal SnSe thermoelectric fibers via laser-induced directional crystallization: From 1D fibers to multidimensional fabrics. *Adv. Mater.* **2020**, *32*, 2002702. [CrossRef]
11. Ruiz-Clavijo, A.; Caballero-Calero, O.; Martín-González, M. Three-dimensional Bi_2Te_3 networks of interconnected nanowires: Synthesis and optimization. *Nanomaterials* **2018**, *8*, 345. [CrossRef]
12. Li, R.-P.; Lu, S.-Y.; Lin, Y.-J.; Chen, C.-Y. Direct Observation of the Epitaxial Growth of Bismuth Telluride Topological Insulators from One-Dimensional Heterostructured Nanowires. *Nanomaterials* **2022**, *12*, 2236. [CrossRef]
13. Zhang, T.; Li, K.; Chen, M.; Wang, Z.; Ma, S.; Zhang, N.; Wei, L. High-performance, flexible, and ultralong crystalline thermoelectric fibers. *Nano Energy* **2017**, *41*, 35–42. [CrossRef]
14. Sun, M.; Tang, G.; Liu, W.; Qian, G.; Huang, K.; Chen, D.; Qian, Q.; Yang, Z. Sn-Se alloy core fibers. *J. Alloys Compd.* **2017**, *725*, 242–247. [CrossRef]
15. Qian, G.; Sun, M.; Tang, G.; Liu, W.; Shi, Z.; Qian, Q.; Zhang, Q.; Yang, Z. High-performance and high-stability bismuth selenide core thermoelectric fibers. *Mater. Lett.* **2018**, *233*, 63–66. [CrossRef]
16. Sun, M.; Qian, Q.; Tang, G.; Liu, W.; Qian, G.; Shi, Z.; Huang, K.; Chen, D.; Xu, S.; Yang, Z. Enhanced thermoelectric properties of polycrystalline Bi_2Te_3 core fibers with preferentially oriented nanosheets. *APL Mater.* **2018**, *6*, 036103. [CrossRef]
17. Sun, M.; Zhang, P.; Li, Q.; Tang, G.; Zhang, T.; Chen, D.; Qian, Q. Enhanced N-Type Bismuth-Telluride-Based Thermoelectric Fibers via Thermal Drawing and Bridgman Annealing. *Materials* **2022**, *15*, 5331. [CrossRef]
18. Zhuang, H.; Pei, J.; Cai, B.; Dong, J.; Hu, H.; Sun, F.; Pan, Y.; Snyder, G.; Li, J. Thermoelectric performance enhancement in BiSbTe alloy by microstructure modulation via cyclic spark plasma sintering with liquid phase. *Adv. Funct. Mater.* **2021**, *31*, 2009681. [CrossRef]
19. Luo, Q.; Tang, G.; Sun, M.; Qian, G.; Shi, Z.; Qian, Q.; Yang, Z. Single crystal tellurium semiconductor core optical fibers. *Opt. Mater. Express* **2020**, *10*, 1072–1082. [CrossRef]
20. Han, B.; Luo, Q.; Zhang, P.; Zhang, T.; Tang, G.; Chen, Z.; Zhang, H.; Zhong, B.; Zeng, Y.; Sun, M.; et al. Multifunctional single-crystal tellurium core multimaterial fiber via thermal drawing and laser recrystallization. *J. Am. Ceram. Soc.* **2022**, *105*, 1640–1647. [CrossRef]

21. Zhao, L.; Zhang, B.-P.; Li, J.-F.; Zhang, H.; Liu, W. Enhanced thermoelectric and mechanical properties in textured n-type Bi_2Te_3 prepared by spark plasma sintering. *Solid State Sci.* **2008**, *10*, 651–658. [CrossRef]
22. Zhao, L.; Zhang, B.-P.; Liu, W.; Zhang, H.; Li, J.-F. Effects of annealing on electrical properties of n-type Bi_2Te_3 fabricated by mechanical alloying and spark plasma sintering. *J. Alloys Compd.* **2009**, *467*, 91–97. [CrossRef]
23. Jacquot, A.; Farag, N.; Jaegle, M.; Bobeth, M.; Schmidt, J.; Ebling, D.; Böttner, H. Thermoelectric properties as a function of electronic band structure and microstructure of textured materials. *J. Electron. Mater.* **2010**, *39*, 1861–1868. [CrossRef]
24. Manzano, C.V.; Abad, B.; Rojo, M.M.; Koh, Y.R.; Hodson, S.L.; Martínez, A.M.L.; Xu, X.; Shakouri, A.; Sands, T.; Borca-Tasciuc, T.; et al. Anisotropic effects on the thermoelectric properties of highly oriented electrodeposited Bi_2Te_3 films. *Sci. Rep.* **2016**, *6*, 1–8. [CrossRef] [PubMed]
25. Kuznetsov, V.L.; Kuznetsova, L.A.; Kaliazin, A.E.; Rowe, D.M. High performance functionally graded and segmented Bi_2Te_3-based materials for thermoelectric power generation. *J. Mater. Sci.* **2002**, *37*, 2893–2897. [CrossRef]

Disclaimer/Publisher's Note: The statements, opinions and data contained in all publications are solely those of the individual author(s) and contributor(s) and not of MDPI and/or the editor(s). MDPI and/or the editor(s) disclaim responsibility for any injury to people or property resulting from any ideas, methods, instructions or products referred to in the content.

Article

Reduced Thermal Conductivity in Nanostructured AgSbTe$_2$ Thermoelectric Material, Obtained by Arc-Melting

Javier Gainza [1,*], Federico Serrano-Sánchez [1], Oscar J. Dura [2], Norbert M. Nemes [3], Jose Luis Martínez [1], María Teresa Fernández-Díaz [4] and José Antonio Alonso [1,*]

[1] Instituto de Ciencia de Materiales de Madrid (ICMM), Consejo Superior de Investigaciones Científicas (CSIC), Sor Juana Inés de la Cruz 3, 28049 Madrid, Spain
[2] Departamento de Física Aplicada, Universidad de Castilla-La Mancha, 13071 Ciudad Real, Spain
[3] Departamento de Física de Materiales, Universidad Complutense de Madrid, 28040 Madrid, Spain
[4] Institut Laue Langevin, BP 156X, 38042 Grenoble, France
* Correspondence: j.gainza@csic.es (J.G.); ja.alonso@icmm.csic.es (J.A.A.)

Abstract: AgSbTe$_2$ intermetallic compound is a promising thermoelectric material. It has also been described as necessary to obtain LAST and TAGS alloys, some of the best performing thermoelectrics of the last decades. Due to the random location of Ag and Sb atoms in the crystal structure, the electronic structure is highly influenced by the atomic ordering of these atoms and makes the accurate determination of the Ag/Sb occupancy of paramount importance. We report on the synthesis of polycrystalline AgSbTe$_2$ by arc-melting, yielding nanostructured dense pellets. SEM images show a conspicuous layered nanostructuration, with a layer thickness of 25–30 nm. Neutron powder diffraction data show that AgSbTe$_2$ crystalizes in the cubic *Pm-3m* space group, with a slight deficiency of Te, probably due to volatilization during the arc-melting process. The transport properties show some anomalies at ~600 K, which can be related to the onset temperature for atomic ordering. The average thermoelectric figure of merit remains around ~0.6 from ~550 up to ~680 K.

Keywords: thermoelectrics; neutron powder diffraction; layered nanostructuration; thermal conductivity

Citation: Gainza, J.; Serrano-Sánchez, F.; Dura, O.J.; Nemes, N.M.; Martínez, J.L.; Fernández-Díaz, M.T.; Alonso, J.A. Reduced Thermal Conductivity in Nanostructured AgSbTe$_2$ Thermoelectric Material, Obtained by Arc-Melting. *Nanomaterials* **2022**, *12*, 3910. https://doi.org/10.3390/nano12213910

Academic Editors: Zhenyang Wang, Shudong Zhang and Nian Li

Received: 10 October 2022
Accepted: 4 November 2022
Published: 5 November 2022

Publisher's Note: MDPI stays neutral with regard to jurisdictional claims in published maps and institutional affiliations.

Copyright: © 2022 by the authors. Licensee MDPI, Basel, Switzerland. This article is an open access article distributed under the terms and conditions of the Creative Commons Attribution (CC BY) license (https://creativecommons.org/licenses/by/4.0/).

1. Introduction

The technological progress that humanity has witnessed in recent years has also led to an unstoppable increase in energy demand worldwide. In addition, society's dependence on fossil fuels is a cause for concern, so renewable energy sources are becoming increasingly important. Thermoelectric materials, which can directly generate electricity from a temperature gradient, can be a key piece in the near future. These devices have several advantages, such as their reliability and absence of mobile parts as well as their environmental benignity. The efficiency of these materials is assessed by the thermoelectric figure of merit [1,2], *ZT*, a dimensionless parameter that is defined as $ZT = (S^2 \cdot \sigma / \kappa) \cdot T$, where S is the Seebeck coefficient, σ is the electrical conductivity, κ is the total thermal conductivity (which is the sum of the lattice and electronic contributions) and T is the absolute temperature.

Among the best thermoelectric materials, tellurides are perhaps the most studied compounds [3–6]. The most prominent, lead telluride (PbTe), has been widely used in the past century [7], and even today remains one of the best performing thermoelectrics [8–11]. Many approaches have been made in the past in order to enhance the thermoelectric performance of PbTe, such as the use of the band convergence concept [12] or deep defect level engineering [13], the implementation of solid solutions [14], the application of lattice strains [15], the introduction of different discordant atoms [8,16,17] and the design of all-scale hierarchical architectures [18]. One of the strategies that has been proven useful is the mixture of PbTe with AgSbTe$_2$, resulting in a compound known as LAST-18 [11,19–21] with a remarkably good thermoelectric performance [11,22]. These kinds of chalcogenide compounds are the paradigmatic example of materials with a low thermal conductivity, due to the high bond anharmonicity [23,24], caused by the lone-pair electrons present in the structure [25].

Wernick and Benson synthesized $AgSbTe_2$ for the first time in 1957 [26], and it has been well known in the thermoelectric community since then [22,27,28]. It was defined as a narrow band gap semiconductor with a rock-salt type structure where Ag and Sb atoms are located at random in the cationic sublattice [29]. The partial atomic ordering of these Ag/Sb atoms in the $AgSbTe_2$ compound is an important feature when analyzing this material, since it strongly influences the electronic structure near the Fermi level [29]. Therefore, the precise determination of the Ag/Sb occupancy in this telluride has a significant relevance in terms of transport properties. This cation disordering is often found in other ternary cubic chalcogenide compounds, with the associated point defects usually leading to a poor reproducibility of the properties of these materials [30]. Recently, Roychowdhury et al. have proven that the atomic ordering can be achieved in $AgSbTe_2$ with cadmium (Cd) doping [31]. This opens a new paradigm, since up until then, most thermoelectric materials were optimized by adding disorder [32].

The LAST compound $AgPb_mSbTe_{2+m}$ also exhibits the honor to be the bulk material in which nanostructuring was first reported [21,22,33–35]. Since then, this approach has been widely used in different thermoelectric compounds with the aim to improve the thermoelectric figure of merit, ZT [36–41]. Nanostructuring can produce, for instance, an increase in the Seebeck coefficient, by means of an increased quantization of the density of states [42], or a reduction in the lattice thermal conductivity, by means of an enhanced phonon scattering mechanism [43]. $AgSbTe_2$ is reported to show a natural formation of nanoscale domains with different orderings on the cation sublattice [44], which is an example of increased phonon scattering, due to a nanostructuring effect.

We have previously used the arc-melting technique to synthesize different chalcogenides and nanostructured compounds, such as PbTe [45], GeTe [46–48], Bi_2Te_3 [49–51] or SnSe [39,52–58]. This synthesis method has the advantage to be very fast, compared to other techniques; the reaction itself happens in only a few seconds, and the entire process can be completed in several minutes. Furthermore, the sample is obtained in the form of dense ingots, which is a useful outcome when we think about the possibility to scale this process up. Here, we report on the synthesis and characterization of the ternary compound $AgSbTe_2$ synthesized by arc-melting. Using this fast and straightforward technique, we can obtain highly dense pellets with nanostructuration in the layers, easily observed by scanning electron microscopy (SEM). The arc-melted compound has been studied by means of X-ray diffraction (XRD) and neutron powder diffraction (NPD), and we have measured its main thermoelectric properties; electrical resistivity, Seebeck coefficient and thermal conductivity, to gain some knowledge about its thermoelectric performance at high temperature.

2. Materials and Methods

$AgSbTe_2$ was synthesized in an Edmund Buhler MAM-1 mini-arc furnace. The pressed pellet was placed in a water-cooled copper crucible, and was melted by a voltaic arc, created by a tungsten electrode in an inert Ar atmosphere. The melting process was repeated three times to ensure the homogeneity of the sample. The reagents were pure elements of Ag (99.9%, Goodfellow Metals, Cambridgeshire, UK), Sb (99.5%, Alfa Aesar, Haverhill, MA, USA) and Te (99.99%, Alfa Aesar, Haverhill, MA, USA), which were weighted (~1.5 g) and mixed, according to the stoichiometric ratio. A small part of the resulting ingot was cut and ground to powder to perform the structural characterization, and the rest of the sample was cold pressed in a Retsch (Haan, Germany) Pellet Press PP25 under an isostatic pressure of 10 MPa to do the transport measurements. This final pellet is typically ~10 mm in diameter and ~2 mm in thickness. The density of the cold-pressed pellet was ~90% of the theoretical crystallographic density. The high-temperature Seebeck coefficient was measured using an MMR technologies instrument under vacuum (10^3 mbar) from room temperature up to ~750 K. Conventional van der Pauw geometry was employed to determine the electrical resistivity. The total thermal conductivity was calculated from the thermal diffusivity (α)

using a Linseis LFA 1000 equipment, by the laser-flash technique. The thermal conductivity (κ) is determined from $\kappa = \alpha \cdot c_p \cdot d$, where c_p is the specific heat, calculated using the Dulong–Petit equation, and d is the sample density.

Phase characterization was carried out for the pulverized sample using X-Ray diffraction (XRD) on a Bruker-AXS D8 (Karlsruhe, Germany) diffractometer run by DIFFRACT-PLUS software (version 2.5.0, Bruker, Karlsruhe, Germany) in Bragg–Brentano reflection geometry with Cu Kα radiation (λ = 1.5418 Å). Furthermore, the NPD was used to characterize the crystal structure in detail. High-resolution patterns were collected in the D2B diffractometer at the Institut Laue-Langevin, Grenoble, France, in the high-flux configuration with a neutron wavelength λ = 1.549 Å, at 298 K. Around 2 g of the sample were measured in a vanadium can. The diffraction data were analyzed using the Rietveld method, employing the FULLPROF program (version Sept. 2018, Institut Laue-Langevin, Grenoble, France). The coherent scattering lengths of Ag, Sb and Te used in the refinement were 5.92, 5.57 and 5.80 fm, respectively. The profile parameters included in the refinement were the background as a set of refinable points, peak shape, asymmetry and FWHM parameters. The structural parameters included the scale factor, lattice parameters, atomic positions, isotropic atomic displacement parameters and occupancy factors. Scanning electron microscopy (SEM) images of an as-grown pellet were collected with a table-top Hitachi TM-1000 microscope (Hitachi, Japan).

3. Results & Discussion

3.1. Crystallographic Analysis by the XRD and NPD

AgSbTe$_2$ was obtained as a well-crystallized sample with negligible impurities (Figure 1). The laboratory XRD patterns display the expected AgSbTe$_2$ cubic phase, defined in the space group $Pm\overline{3}m$ with the lattice parameter a = 6.0788 Å. In Figure 1a, the peaks appear indexed in the mentioned cubic lattice. There are no additional reflections that could suggest a superstructure or a different space group. The pattern displays a slight preferred orientation effect and minor impurities of Ag$_2$Te and Sb$_2$Te$_3$, as expected, according to previous literature reports [59].

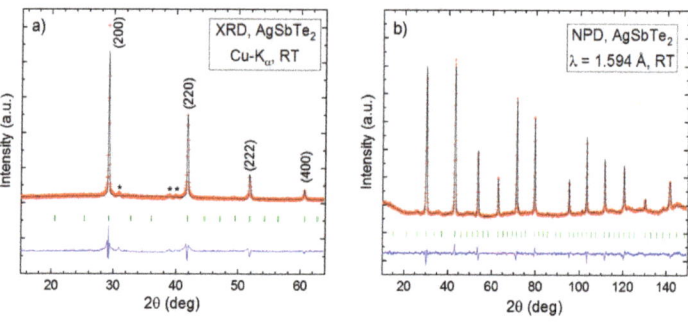

Figure 1. (a) XRD and (b) NPD patterns of AgSbTe$_2$ at room temperature. The experimental points are shown in red, the calculated model in black and the difference in blue. The star denotes the known impurities detected in the X-ray pattern.

A detailed structural investigation was performed by NPD at 295 K, which was essential as a bulk analysis to remove any orientation effects and to precisely determine the atomic displacement parameters (ADPs). The Rietveld refinement was performed using the CsCl-type structure, defined in the $Pm\overline{3}m$ space group, which shows a good agreement with the observed pattern (Figure 1b). Previous reports have also proposed the F-centered $Fm\overline{3}m$ space group to define the AgSbTe$_2$ structure, but we could not improve the refinement of the primitive unit cell. Moreover, a structural description from the single-crystal diffraction data discarded a type-F lattice [60]. Thus, at $Pm\overline{3}m$, Ag and Sb atoms share randomly 3c

($\frac{1}{2}$, $\frac{1}{2}$, 0) Wyckoff positions, Sb is additionally located at 1b ($\frac{1}{2}$, $\frac{1}{2}$, $\frac{1}{2}$) and Te at 1a (0, 0, 0) and 3d (0, $\frac{1}{2}$, 0) Wyckoff positions. Table 1 lists the experimental set-up and Table 2, the determined structural parameters.

Table 1. NPD experimental parameters of AgSbTe$_2$ at room temperature.

Diffraction Parameters	
Wavelength (Å)	1.594
2θ range (°)	0.07–159.97
2θ step (°)	0.05
Temperature (K)	295
Rietveld Refinement	3199 data points
No. of Parameters	67
Structural parameters	
Formula	AgSbTe$_2$
Space Group	$Pm\bar{3}m$
Z	2
a (Å)	6.0788(1)
V(Å^3)	224.619(7)
Theoretical Density (g·cm^{-3})	7.168

Table 2. Structural parameters obtained from the refinement of the NPD data of AgSbTe$_2$ at room temperature.

Atomic Parameters						
Atom	Wyckoff site	x	y	z	U_{eq} (Å^2)	Occ. (<1)
Te1	3d	0.00000	0.50000	0.00000	0.027 (7)	0.82 (3)
Te2	1a	0.00000	0.00000	0.00000	0.014 (6)	1.0 (1)
Sb1	1b	0.50000	0.50000	0.50000	0.038 (8)	1.0 (1)
Ag	3c	0.50000	0.50000	0.00000	0.026 (5)	0.6667
Sb2	3c	0.50000	0.50000	0.00000	0.026 (5)	0.3333
Atomic Displacement Parameters (Å^2)						
		U^{11}	U^{22}	U^{33}	U^{12}	U^{13}
Te1		0.022 (5)	0.04 (1)	0.022 (5)	0.00000	0.00000
Sb		0.038 (8)	0.038 (8)	0.038 (8)	0.00000	0.00000
Ag		0.036 (5)	0.036 (5)	0.005 (6)	0.00000	0.00000
Sb2		0.036 (5)	0.036 (5)	0.005 (6)	0.00000	0.00000
Agreement Factors						Bond Distance (Å)
R_I(%)		R_p(%)	R_{wp}(%)	R_{exp}(%)	χ^2	d(Ag/Sb-Te)
5.2		3.2	4.1	2.8	2.1	3.03939(6)

A slight Te deficiency at the 3d position is observed, which agrees with some Te volatilization during the arc-melting process, and these anion vacancies will increase the Fermi level and electron concentration. Figure 2 illustrates a view of the $Pm\bar{3}m$ crystal structure, with the anisotropic atomic displacement ellipsoids (ADPs) for the 3c and 3d sites. This is a rock-salt-like structure, with Te and Sb as contiguous atoms and SbX$_6$

(X = Ag, Sb) and TeTe$_2$X$_4$ octahedra. It is noteworthy that this model provides only one bonding distance, equivalent for every position. The highly anisotropic ADPs reflect the strong anharmonic bonding, which in turn has been ascribed to the lone s^2 pair effect of Sb atoms [25]. The flat ellipsoids are perpendicular to the Sb-X bond and the elongated 3d-Te ellipsoids suggest more labile Te-Te, Ag/Sb-Te interactions and stronger Ag/Sb-Sb bonds, as well as repulsion effects with the origin at the 1b-Sb position.

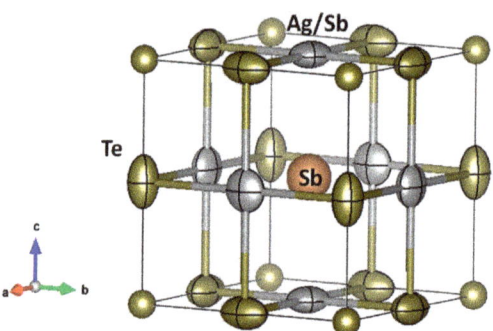

Figure 2. Crystal structure of AgSbTe$_2$, defined in the $Pm\overline{3}m$ space group. The characteristic shape of the ellipsoids (elongated for Te1 and flattened for Ag/Sb2) are highlighted.

The crystal structure of AgSbTe$_2$ is peculiar in the sense that Ag and Sb may be totally disordered, i.e., statistically distributed over the same crystallographic positions, as described in the cubic space group $Fm\overline{3}m$, or partially disordered, as described in the $Pm\overline{3}m$ space group, and found in the present case, where part of Sb and Ag are still distributed at random over the 3c Wyckoff sites (see Table 2, Figure 2). Some authors [31] have found that, by means of Cd doping, the structure tends to increase the ordering; then Ag and Sb(Cd) become ordered by forming nanoregions. The fact that there are no thermal events in the DSC curves of AgSbTe$_2$, as reported by Roychowdhury et al. [61], seems to suggest that AgSbTe$_2$ remains in a partly ordered structure ($Pm\overline{3}m$), although the presence of partially ordered regions, at the nanoscale, that do not give rise to the superstructure reflections at intermediate temperatures, should not be discarded [29].

3.2. Scanning Electron Microscopy

The microstructure of the as-grown AgSbTe$_2$ ingots has been investigated by high-resolution SEM, recorded in a table-top microscope. Some selected micrographs are shown in Figure 3. The material seems to be quite homogeneous, while it consists of a stacking of sheets, each of them presumably single-crystalline, with the large surfaces perpendicular to a crystallographic axis. This stacking of sheets seems to be a consequence of the arc-melting synthesis procedure, probably due to the inherent fast cooling protocol. Figure 3c,d assess a layer thickness in the 25–30 nm range. This strong nanostructuration in the layers, accounting for the ease of the cleavage of this material, is also responsible for the observed decrease of the thermal conductivity, as described below.

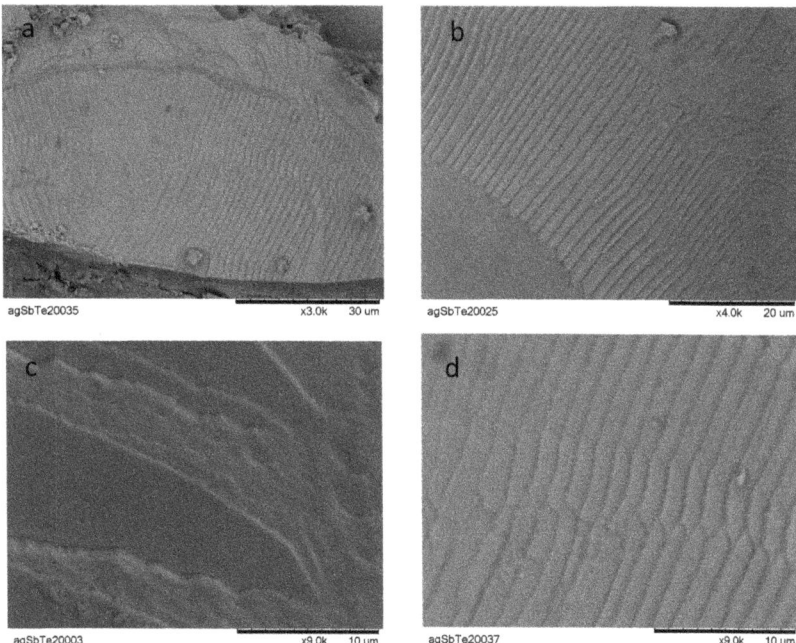

Figure 3. SEM micrographs with (**a**) 3000×, (**b**) 4000×, (**c**,**d**) 9000× magnification, illustrating the nanostructuration in the layers observed in this material, grown from arc melting.

3.3. Thermoelectric Properties

Figure 4 displays the electrical transport properties of AgSbTe$_2$. The resistivity (Figure 4a) shows almost constant values with the temperature, with a small variation from 2.6×10^{-4} to 2.8×10^{-4} $\Omega \cdot$m in the temperature range 300–770 K. It shows increasing values up to 550 K, with a small bump that has been related to the cation disorder at high temperatures [61], and which will also be apparent as a peak in the weighted, as we will discuss later. This is followed by a slight decrease in the resistivity, most likely a result of a minority carriers' excitation. These values are above those that have been previously reported, within the range 0.5–2×10^{-4} $\Omega \cdot$m for pristine AgSbTe$_2$, prepared by melt-quench or mechanical alloying and SPS [62,63]. The Seebeck coefficient evolution with the temperature (Figure 4b) exhibits a steady increase from 200 µV K^{-1} to 340 V K^{-1} up to 540 K, a plateau up to 700 K and a stark reduction above 700 K, due to the bipolar contribution. This behavior is found in other AgSbTe$_2$ samples, with a maximum value and temperature determined by the contribution of both carrier types in a semimetal as [64]:

$$S = \frac{S_p \sigma_p + S_e \sigma_e}{\sigma_p + \sigma_e}$$

AgSbTe$_2$ displays a high Seebeck coefficient, due to its valley degeneracy and hole heavy effective band mass. Here, the samples display an increase of the Seebeck values above 300 µV·K^{-1}, higher than those reported on the literature [62,63], possibly due to the lower values of the carrier concentration, which would also match the increased resistivity. This is opposite to the results obtained by the high-pressure high-temperature preparation [65].

The power factor (Figure 4c) follows the Seebeck coefficient evolution with maximum values close to 0.4 mW·m^{-1}·K^{-2}, in the range from 540 to 640 K [31,62,63]. These values are comparable to those reported in the literature (0.5–0.8), slightly limited by the higher electrical resistivity. AgSbTe$_2$ presents a melting point slightly above ~800 K, and thus, the measurements have been performed up to ~760 K, to ensure the reproducibility of the results.

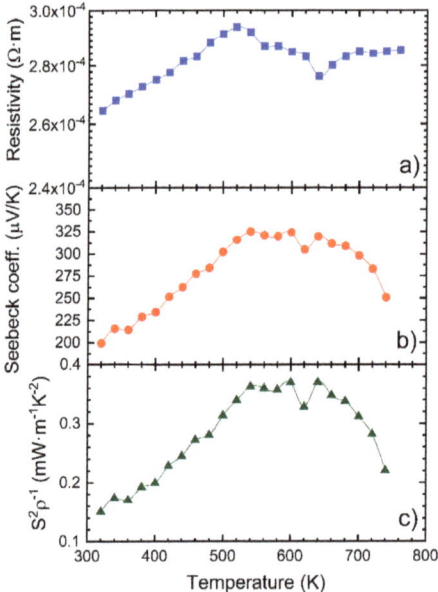

Figure 4. (a) Resistivity, (b) Seebeck coefficient and (c) power factor of the arc-melted AgSbTe$_2$. The power factor is calculated from the experimental resistivity and the Seebeck coefficient.

3.4. Thermal Conductivity

The total thermal conductivity vs. the temperature curve is shown in Figure 5a. It shows an almost constant value close to 0.4 W m^{-1} K^{-1}, below the previously reported values for pristine AgSbTe$_2$ [31]. The electronic and lattice contributions were determined by the Wiedemann–Franz law. The lattice thermal conductivity values are rather close to those of the total conductivity, due to a small electronic contribution (Figure 5b). AgSbTe$_2$ exhibits an extremely low intrinsic thermal conductivity, which has been related to the anharmonic bonding, due to the lone pair effect [25]; it additionally presents a spontaneous nanostructuration, due to the different cationic ordering in the nanoscale [44,65,66]. Our specimen, prepared by arc melting, has even lower values than those reported in the literature, reaching down to 0.32 W·m^{-1} K^{-1} at 623 K, while previous reports show minimum values of 0.4 W·m^{-1} K^{-1}. This enhanced phonon scattering, observed in our samples, is a consequence of the characteristic nanostructuration, obtained after arc-melting of the samples, as observed in the SEM images (Figure 3), and it was described in other thermoelectric materials [39,49,52].

The minimum value of the lattice thermal conductivity, observed at 623 K, followed by a conspicuous increase (Figure 5) can be explained, based on the effect of the atomic ordering happening at that temperature. It has been reported that this atomic ordering effect can be inferred from the unconventional temperature dependence of the transport properties [30], such as in the lattice thermal conductivity, as well as in the resistivity, the Seebeck coefficient and the weighted mobility.

In Figure 6a, the weighted mobility dependence on the temperature is shown. It increases up to 520–550 K, when the cationic disorder at a high temperature increases the carrier scattering. These values are much closer to those reported for the hole-mobility (~15 cm^2 V^{-1} s^{-1}), while much lower than those found for the electron mobility (~10^4 cm^2 V^{-1} s^{-1}), as the heavy p-type band is the main contribution to the Seebeck coefficient. The maximum weighted mobility is consistent with the temperature evolution of the Seebeck coefficient and the resistivity, showing the significance of the cationic ordering in the electrical transport properties [31].

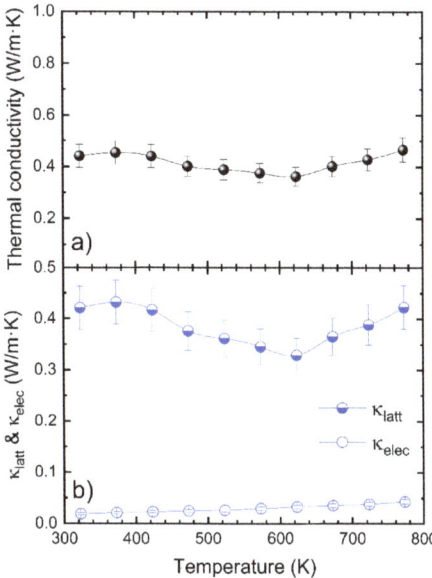

Figure 5. (**a**) Total and (**b**) lattice and electronic contributions to the thermal conductivity as a function of temperature for the AgSbTe$_2$ compound. There is an increase in the lattice thermal conductivity at 623 K, probably related to the effect of the atomic ordering happening around that temperature range.

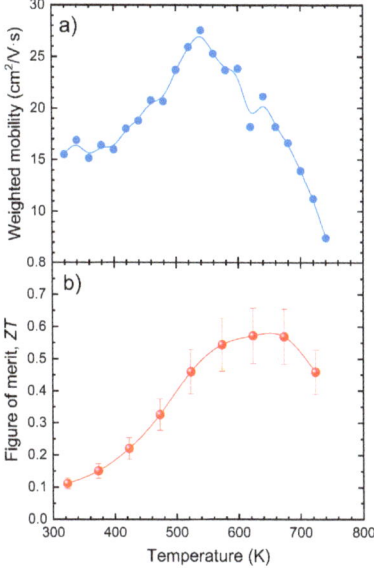

Figure 6. (**a**) Weighted mobility and (**b**) thermoelectric figure of the merit for the arc-melted AgSbTe$_2$ compound.

Overall, the figure of merit (Figure 6b) reaches a non-negligible value of 0.6 at 680 K, following the increase of the Seebeck coefficient, despite the reduced electrical conductivity.

Nevertheless, these samples display the effect of an even more reduced thermal conductivity, as obtained by the nanostructuration of the arc-melted samples. The experimental data for this arc-melted compound are shown in Figure 7, together with other reported data for similar compositions, for the sake of comparison. Owing to this extremely low thermal conductivity, arc-melted samples are a suitable platform for the optimization of the electrical transport properties. Moreover, this synthesis procedure presents advantages over those previously described, since in a single step we obtain, by arc melting, a material in the form of an ingot, which can be directly implemented into a thermoelectric device, without requiring additional (and expensive) treatments, such as SPS, hot pressing, etc.

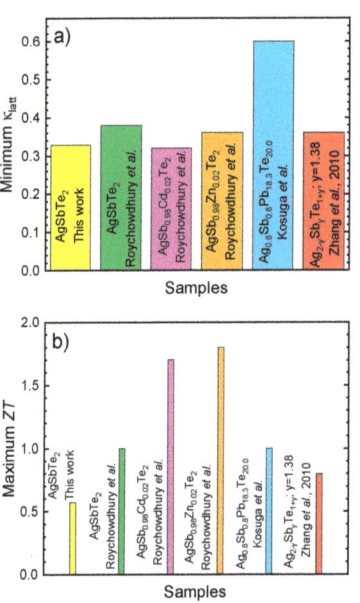

Figure 7. (**a**) Maximum thermoelectric figure of merit, ZT, and (**b**) minimum lattice thermal conductivity for several tellurides [31,61,67,68], compared with the composition analyzed in this work, $AgSbTe_2$.

4. Conclusions

We have prepared the telluride $AgSbTe_2$ through the arc-melting technique, yielding a nanostructured dense pellet. The SEM images reveal a conspicuous layered nanostructuration, with layer thicknesses of 25–30 nm. The refinement of the crystal structure from the neutron powder diffraction data at RT has performed well, considering a $Pm\overline{3}m$ space group and it reveals a tellurium deficiency that can be associated with the volatilization during the arc-melting process. This structure involves an intrinsic partial disordering of Ag/Sb, statistically distributed over the $3c$ Wyckoff positions. The possible atomic ordering reported for this compound can be inferred by an unconventional temperature dependence of the transport properties at high temperatures; this event can be detected at around ~600 K, when the weighted mobility and the lattice thermal conductivity, for example, show an anomaly in their behavior. The average thermoelectric figure of merit of this arc-melted compound remains at around ~0.6 from ~550 up to ~680 K, an important parameter to bear in mind to implement this material in practical devices.

Author Contributions: Conceptualization, J.G. and J.A.A.; methodology, J.G., O.J.D. and M.T.F.-D.; software, J.G., F.S.-S. and J.A.A.; validation, N.M.N. and J.A.A.; formal analysis, J.G. and F.S.-S.; investigation, J.G., F.S.-S. and J.A.A.; resources, O.J.D. and M.T.F.-D.; data curation, J.G. and F.S.-S.; writing—original draft preparation, J.G. and F.S.-S.; writing—review and editing, N.M.N., J.L.M. and J.A.A.; visualization, J.G., F.S.-S. and J.A.A.; supervision, J.L.M. and J.A.A.; project administration,

J.L.M. and J.A.A.; funding acquisition, J.L.M. and J.A.A. All authors have read and agreed to the published version of the manuscript.

Funding: This research was funded by the Spanish Ministry for Science and Innovation (MCIN/AEI/ 10.13039/501100011033) with the grant numbers PID2021-122477OB-I00 and TED2021-129254B-C22.

Institutional Review Board Statement: Not applicable.

Informed Consent Statement: Not applicable.

Data Availability Statement: Experimental raw data are available to the reader from lead contact upon reasonable request.

Acknowledgments: We thank the Spanish Ministry for Science and Innovation (MCIN/AEI/10.13039/ 501100011033) for their generous funding, with grant numbers PID2021-122477OB-I00 and TED2021- 129254B-C22. JG thanks MICINN for granting the contract PRE2018-083398. The authors wish to express their gratitude to the ILL technical staff for making the facilities available for the neutron powder diffraction experiment.

Conflicts of Interest: The authors declare no conflict of interest.

References

1. Snyder, G.J.; Toberer, E.S. Complex Thermoelectric Materials. *Nat. Mater.* **2008**, *7*, 105–114. [CrossRef]
2. Shi, X.-L.; Zou, J.; Chen, Z.-G. Advanced Thermoelectric Design: From Materials and Structures to Devices. *Chem. Rev.* **2020**, *120*, 7399–7515. [CrossRef] [PubMed]
3. Witting, I.T.; Chasapis, T.C.; Ricci, F.; Peters, M.; Heinz, N.A.; Hautier, G.; Snyder, G.J. The Thermoelectric Properties of Bismuth Telluride. *Adv. Electron. Mater.* **2019**, *5*, 1800904. [CrossRef]
4. Liu, M.; Zhu, J.; Cui, B.; Guo, F.; Liu, Z.; Zhu, Y.; Guo, M.; Sun, Y.; Zhang, Q.; Zhang, Y.; et al. High-Performance Lead-Free Cubic GeTe-Based Thermoelectric Alloy. *Cell Rep. Phys. Sci.* **2022**, *3*, 100902. [CrossRef]
5. Tan, G.; Shi, F.; Hao, S.; Chi, H.; Zhao, L.-D.; Uher, C.; Wolverton, C.; Dravid, V.P.; Kanatzidis, M.G. Codoping in SnTe: Enhancement of Thermoelectric Performance through Synergy of Resonance Levels and Band Convergence. *J. Am. Chem. Soc.* **2015**, *137*, 5100–5112. [CrossRef] [PubMed]
6. Lin, S.; Li, W.; Chen, Z.; Shen, J.; Ge, B.; Pei, Y. Tellurium as a High-Performance Elemental Thermoelectric. *Nat. Commun.* **2016**, *7*, 10287. [CrossRef] [PubMed]
7. LaLonde, A.D.; Pei, Y.; Wang, H.; Jeffrey Snyder, G. Lead Telluride Alloy Thermoelectrics. *Mater. Today* **2011**, *14*, 526–532. [CrossRef]
8. Luo, Z.-Z.; Cai, S.; Hao, S.; Bailey, T.P.; Luo, Y.; Luo, W.; Yu, Y.; Uher, C.; Wolverton, C.; Dravid, V.P.; et al. Extraordinary Role of Zn in Enhancing Thermoelectric Performance of Ga-Doped n-Type PbTe. *Energy Environ. Sci.* **2022**, *15*, 368–375. [CrossRef]
9. Brod, M.K.; Toriyama, M.Y.; Snyder, G.J. Orbital Chemistry That Leads to High Valley Degeneracy in PbTe. *Chem. Mater.* **2020**, *32*, 9771–9779. [CrossRef]
10. Tan, G.; Zhang, X.; Hao, S.; Chi, H.; Bailey, T.P.; Su, X.; Uher, C.; Dravid, V.P.; Wolverton, C.; Kanatzidis, M.G. Enhanced Density-of-States Effective Mass and Strained Endotaxial Nanostructures in Sb-Doped $Pb_{0.97}Cd_{0.03}$Te Thermoelectric Alloys. *ACS Appl. Mater. Interfaces* **2019**, *11*, 9197–9204. [CrossRef]
11. Pei, Y.; Gibbs, Z.M.; Gloskovskii, A.; Balke, B.; Zeier, W.G.; Snyder, G.J. Optimum Carrier Concentration in N-Type PbTe Thermoelectrics. *Adv. Energy Mater.* **2014**, *4*, 1400486. [CrossRef]
12. Pei, Y.; Shi, X.; LaLonde, A.; Wang, H.; Chen, L.; Snyder, G.J. Convergence of Electronic Bands for High Performance Bulk Thermoelectrics. *Nature* **2011**, *473*, 66–69. [CrossRef] [PubMed]
13. Zhang, Q.; Song, Q.; Wang, X.; Sun, J.; Zhu, Q.; Dahal, K.; Lin, X.; Cao, F.; Zhou, J.; Chen, S.; et al. Deep Defect Level Engineering: A Strategy of Optimizing the Carrier Concentration for High Thermoelectric Performance. *Energy Environ. Sci.* **2018**, *11*, 933–940. [CrossRef]
14. Wang, X.-K.; Veremchuk, I.; Bobnar, M.; Zhao, J.-T.; Grin, Y. Solid Solution $Pb_{1-x}Eu_xTe$: Constitution and Thermoelectric Behavior. *Inorg. Chem. Front.* **2016**, *3*, 1152–1159. [CrossRef]
15. Wu, Y.; Chen, Z.; Nan, P.; Xiong, F.; Lin, S.; Zhang, X.; Chen, Y.; Chen, L.; Ge, B.; Pei, Y. Lattice Strain Advances Thermoelectrics. *Joule* **2019**, *3*, 1276–1288. [CrossRef]
16. Luo, Z.-Z.; Cai, S.; Hao, S.; Bailey, T.P.; Su, X.; Spanopoulos, I.; Hadar, I.; Tan, G.; Luo, Y.; Xu, J.; et al. High Figure of Merit in Gallium-Doped Nanostructured n-Type PbTe-xGeTe with Midgap States. *J. Am. Chem. Soc.* **2019**, *141*, 16169–16177. [CrossRef]
17. Dutta, M.; Biswas, R.K.; Pati, S.K.; Biswas, K. Discordant Gd and Electronic Band Flattening Synergistically Induce High Thermoelectric Performance in N-Type PbTe. *ACS Energy Lett.* **2021**, *6*, 1625–1632. [CrossRef]
18. Zhao, L.D.; Wu, H.J.; Hao, S.Q.; Wu, C.I.; Zhou, X.Y.; Biswas, K.; He, J.Q.; Hogan, T.P.; Uher, C.; Wolverton, C.; et al. All-Scale Hierarchical Thermoelectrics: MgTe in PbTe Facilitates Valence Band Convergence and Suppresses Bipolar Thermal Transport for High Performance. *Energy Environ. Sci.* **2013**, *6*, 3346–3355. [CrossRef]

19. Pei, Y.; Lensch-Falk, J.; Toberer, E.S.; Medlin, D.L.; Snyder, G.J. High Thermoelectric Performance in PbTe Due to Large Nanoscale Ag$_2$Te Precipitates and La Doping. *Adv. Funct. Mater.* **2011**, *21*, 241–249. [CrossRef]
20. Levin, E.M.; Cook, B.A.; Ahn, K.; Kanatzidis, M.G.; Schmidt-Rohr, K. Electronic Inhomogeneity and Ag:Sb Imbalance of Ag1-y Pb18Sb1+zTe20 High-Performance Thermoelectrics Elucidated by 125Te and 207Pb NMR. *Phys. Rev. B* **2009**, *80*, 115211. [CrossRef]
21. Perlt, S.; Höche, T.; Dadda, J.; Müller, E.; Bauer Pereira, P.; Hermann, R.; Sarahan, M.; Pippel, E.; Brydson, R. Microstructure Analyses and Thermoelectric Properties of Ag$_{1-x}$Pb$_{18}$Sb$_{1+y}$Te$_{20}$. *J. Solid State Chem.* **2012**, *193*, 58–63. [CrossRef]
22. Hsu, K.F.; Loo, S.; Guo, F.; Chen, W.; Dyck, J.S.; Uher, C.; Hogan, T.; Polychroniadis, E.K.; Kanatzidis, M.G. Cubic AgPbmSbTe2+m: Bulk Thermoelectric Materials with High Figure of Merit. *Science* **2004**, *303*, 818–821. [CrossRef] [PubMed]
23. Heremans, J.P. The Anharmonicity Blacksmith. *Nat. Phys.* **2015**, *11*, 990–991. [CrossRef]
24. Chang, C.; Zhao, L.-D. Anharmoncity and Low Thermal Conductivity in Thermoelectrics. *Mater. Today Phys.* **2018**, *4*, 50–57. [CrossRef]
25. Nielsen, M.D.; Ozolins, V.; Heremans, J.P. Lone Pair Electrons Minimize Lattice Thermal Conductivity. *Energy Environ. Sci.* **2013**, *6*, 570–578. [CrossRef]
26. Wernick, J.H.; Benson, K.E. New Semiconducting Ternary Compounds. *J. Phys. Chem. Solids* **1957**, *3*, 157–159. [CrossRef]
27. Hockings, E.F. The Thermal Conductivity of Silver Antimony Telluride. *J. Phys. Chem. Solids* **1959**, *10*, 341–342. [CrossRef]
28. Irie, T.; Takahama, T.; Ono, T. The Thermoelectric Properties of AgSbTe$_2$–AgBiTe$_2$, AgSbTe$_2$–PbTe and–SnTe Systems. *Jpn. J. Appl. Phys.* **1963**, *2*, 72–82. [CrossRef]
29. Ghosh, T.; Roychowdhury, S.; Dutta, M.; Biswas, K. High-Performance Thermoelectric Energy Conversion: A Tale of Atomic Ordering in AgSbTe$_2$. *ACS Energy Lett.* **2021**, *6*, 2825–2837. [CrossRef]
30. Jang, H.; Toriyama, M.Y.; Abbey, S.; Frimpong, B.; Male, J.P.; Snyder, G.J.; Jung, Y.S.; Oh, M. Suppressing Charged Cation Antisites via Se Vapor Annealing Enables P-Type Dopability in AgBiSe$_2$–SnSe Thermoelectrics. *Adv. Mater.* **2022**, *34*, 2204132. [CrossRef]
31. Roychowdhury, S.; Ghosh, T.; Arora, R.; Samanta, M.; Xie, L.; Singh, N.K.; Soni, A.; He, J.; Waghmare, U.V.; Biswas, K. Enhanced Atomic Ordering Leads to High Thermoelectric Performance in AgSbTe$_2$. *Science* **2021**, *371*, 722–727. [CrossRef] [PubMed]
32. Liu, Y.; Ibáñez, M. Tidying up the Mess. *Science* **2021**, *371*, 678–679. [CrossRef] [PubMed]
33. Korkosz, R.J.; Chasapis, T.C.; Lo, S.; Doak, J.W.; Kim, Y.J.; Wu, C.-I.; Hatzikraniotis, E.; Hogan, T.P.; Seidman, D.N.; Wolverton, C.; et al. High ZT in p-Type (PbTe)$_{1-2x}$(PbSe)$_x$(PbS)$_x$ Thermoelectric Materials. *J. Am. Chem. Soc.* **2014**, *136*, 3225–3237. [CrossRef] [PubMed]
34. Cook, B.A.; Kramer, M.J.; Harringa, J.L.; Han, M.-K.; Chung, D.-Y.; Kanatzidis, M.G. Analysis of Nanostructuring in High Figure-of-Merit Ag$_{1-x}$PbmSbTe$_{2+m}$ Thermoelectric Materials. *Adv. Funct. Mater.* **2009**, *19*, 1254–1259. [CrossRef]
35. Kanatzidis, M.G. Nanostructured Thermoelectrics: The New Paradigm? *Chem. Mater.* **2010**, *22*, 648–659. [CrossRef]
36. Toprak, M.S.; Stiewe, C.; Platzek, D.; Williams, S.; Bertini, L.; Müller, E.; Gatti, C.; Zhang, Y.; Rowe, M.; Muhammed, M. The Impact of Nanostructuring on the Thermal Conductivity of Thermoelectric CoSb$_3$. *Adv. Funct. Mater.* **2004**, *14*, 1189–1196. [CrossRef]
37. Tan, G.; Zhao, L.-D.; Shi, F.; Doak, J.W.; Lo, S.-H.; Sun, H.; Wolverton, C.; Dravid, V.P.; Uher, C.; Kanatzidis, M.G. High Thermoelectric Performance of P-Type SnTe via a Synergistic Band Engineering and Nanostructuring Approach. *J. Am. Chem. Soc.* **2014**, *136*, 7006–7017. [CrossRef] [PubMed]
38. Zhu, Y.; Carrete, J.; Meng, Q.-L.; Huang, Z.; Mingo, N.; Jiang, P.; Bao, X. Independently Tuning the Power Factor and Thermal Conductivity of SnSe via Ag$_2$S Addition and Nanostructuring. *J. Mater. Chem. A Mater.* **2018**, *6*, 7959–7966. [CrossRef]
39. Gainza, J.; Serrano-Sánchez, F.; Gharsallah, M.; Carrascoso, F.; Bermúdez, J.; Dura, O.J.; Mompean, F.J.; Biskup, N.; Meléndez, J.J.; Martínez, J.L.; et al. Evidence of Nanostructuring and Reduced Thermal Conductivity in N-Type Sb-Alloyed SnSe Thermoelectric Polycrystals. *J. Appl. Phys.* **2019**, *126*, 045105. [CrossRef]
40. Luo, Y.; Cai, S.; Hua, X.; Chen, H.; Liang, Q.; Du, C.; Zheng, Y.; Shen, J.; Xu, J.; Wolverton, C.; et al. High Thermoelectric Performance in Polycrystalline SnSe Via Dual-Doping with Ag/Na and Nanostructuring With Ag$_8$SnSe$_6$. *Adv. Energy Mater.* **2019**, *9*, 1803072. [CrossRef]
41. Perumal, S.; Bellare, P.; Shenoy, U.S.; Waghmare, U.V.; Biswas, K. Low Thermal Conductivity and High Thermoelectric Performance in Sb and Bi Codoped GeTe: Complementary Effect of Band Convergence and Nanostructuring. *Chem. Mater.* **2017**, *29*, 10426–10435. [CrossRef]
42. Vaqueiro, P.; Powell, A.V. Recent Developments in Nanostructured Materials for High-Performance Thermoelectrics. *J. Mater. Chem.* **2010**, *20*, 9577. [CrossRef]
43. Li, J.-F.; Liu, W.-S.; Zhao, L.-D.; Zhou, M. High-Performance Nanostructured Thermoelectric Materials. *NPG Asia Mater.* **2010**, *2*, 152–158. [CrossRef]
44. Ma, J.; Delaire, O.; May, A.F.; Carlton, C.E.; McGuire, M.A.; VanBebber, L.H.; Abernathy, D.L.; Ehlers, G.; Hong, T.; Huq, A.; et al. Glass-like Phonon Scattering from a Spontaneous Nanostructure in AgSbTe$_2$. *Nat. Nanotechnol.* **2013**, *8*, 445–451. [CrossRef] [PubMed]
45. Gainza, J.; Serrano-Sánchez, F.; Biskup, N.; Nemes, N.M.; Martínez, J.L.; Fernández-Díaz, M.T.; Alonso, J.A. Influence of Nanostructuration on PbTe Alloys Synthesized by Arc-Melting. *Materials* **2019**, *12*, 3783. [CrossRef]
46. Serrano-Sánchez, F.; Funes, M.; Nemes, N.M.; Dura, O.J.; Martínez, J.L.; Prado-Gonjal, J.; Fernández-Díaz, M.T.; Alonso, J.A. Low Lattice Thermal Conductivity in Arc-Melted GeTe with Ge-Deficient Crystal Structure. *Appl. Phys. Lett.* **2018**, *113*, 1–5. [CrossRef]

47. Gainza, J.; Serrano-Sánchez, F.; Nemes, N.M.; Martínez, J.L.; Fernández-Díaz, M.T.; Alonso, J.A. Features of the High-Temperature Structural Evolution of GeTe Thermoelectric Probed by Neutron and Synchrotron Powder Diffraction. *Metals* **2019**, *10*, 48. [CrossRef]
48. Gainza, J.; Serrano-Sánchez, F.; Nemes, N.M.; Dura, O.J.; Martínez, J.L.; Alonso, J.A. Lower Temperature of the Structural Transition, and Thermoelectric Properties in Sn-Substituted GeTe. *Mater. Today Proc.* **2021**, *44*, 3450–3457. [CrossRef]
49. Gharsallah, M.; Serrano-Sánchez, F.; Bermúdez, J.; Nemes, N.M.; Martínez, J.L.; Elhalouani, F.; Alonso, J.A. Nanostructured Bi_2Te_3 Prepared by a Straightforward Arc-Melting Method. *Nanoscale Res. Lett.* **2016**, *11*, 4–10. [CrossRef]
50. Gharsallah, M.; Serrano-Sanchez, F.; Nemes, N.M.; Martinez, J.L.; Alonso, J.A. Influence of Doping and Nanostructuration on N-Type $Bi_2(Te_{0.8}Se_{0.2})_3$ Alloys Synthesized by Arc Melting. *Nanoscale Res. Lett.* **2017**, *12*, 47. [CrossRef]
51. Serrano-Sánchez, F.; Gharsallah, M.; Nemes, N.M.; Biskup, N.; Varela, M.; Martínez, J.L.; Fernández-Díaz, M.T.; Alonso, J.A. Enhanced Figure of Merit in Nanostructured $(Bi, Sb)_2Te_3$ with Optimized Composition, Prepared by a Straightforward Arc-Melting Procedure. *Sci. Rep.* **2017**, *7*, 6277. [CrossRef]
52. Serrano-Sánchez, F.; Gharsallah, M.; Nemes, N.M.; Mompean, F.J.; Martínez, J.L.; Alonso, J.A. Record Seebeck Coefficient and Extremely Low Thermal Conductivity in Nanostructured SnSe. *Appl. Phys. Lett.* **2015**, *106*, 083902. [CrossRef]
53. Serrano-Sánchez, F.; Nemes, N.M.; Dura, O.J.; Fernandez-Diaz, M.T.; Martínez, J.L.; Alonso, J.A. Structural Phase Transition in Polycrystalline SnSe: A Neutron Diffraction Study in Correlation with Thermoelectric Properties. *J. Appl. Cryst.* **2016**, *49*, 2138–2144. [CrossRef]
54. Gainza, J.; Serrano-Sánchez, F.; Rodrigues, J.E.F.S.; Huttel, Y.; Dura, O.J.; Koza, M.M.; Fernández-Díaz, M.T.; Meléndez, J.J.; Márkus, B.G.; Simon, F.; et al. High-Performance n-Type SnSe Thermoelectric Polycrystal Prepared by Arc-Melting. *Cell Rep. Phys. Sci.* **2020**, *1*, 100263. [CrossRef]
55. Gharsallah, M.; Serrano-Sánchez, F.; Nemes, N.M.; Mompeán, F.J.; Martínez, J.L.; Fernández-Díaz, M.T.; Elhalouani, F.; Alonso, J.A. Giant Seebeck Effect in Ge-Doped SnSe. *Sci. Rep.* **2016**, *6*, 26774. [CrossRef]
56. Gainza, J.; Moltó, S.; Serrano-Sánchez, F.; Dura, O.J.; Fernández-Díaz, M.T.; Biškup, N.; Martínez, J.L.; Alonso, J.A.; Nemes, N.M. $SnSe:K_x$ Intermetallic Thermoelectric Polycrystals Prepared by Arc-Melting. *J. Mater. Sci.* **2022**, *57*, 8489–8503. [CrossRef]
57. Serrano-Sánchez, F.; Gharsallah, M.; Nemes, N.M.; Mompeán, F.J.; Martínez, J.L.; Alonso, J.A. Facile Preparation of SnSe Derivatives in Nanostructured Polycrystalline Form by Arc-Melting Synthesis. *Mater. Today Proc.* **2018**, *5*, 10218–10226. [CrossRef]
58. Gainza, J.; Serrano-Sánchez, F.; Gharsallah, M.; Funes, M.; Carrascoso, F.; Nemes, N.M.; Dura, O.J.; Martínez, J.L.; Alonso, J.A. Nanostructured Thermoelectric Chalcogenides. In *Bringing Thermoelectricity into Reality*; IntechOpen: London, UK, 2018. [CrossRef]
59. Medlin, D.L.; Snyder, G.J. Interfaces in Bulk Thermoelectric Materials. *Curr. Opin. Colloid. Interface Sci.* **2009**, *14*, 226–235. [CrossRef]
60. Quarez, E.; Hsu, K.-F.; Pcionek, R.; Frangis, N.; Polychroniadis, E.K.; Kanatzidis, M.G. Nanostructuring, Compositional Fluctuations, and Atomic Ordering in the Thermoelectric Materials $AgPb_mSbTe_{2+m}$. The Myth of Solid Solutions. *J. Am. Chem. Soc.* **2005**, *127*, 9177–9190. [CrossRef]
61. Roychowdhury, S.; Panigrahi, R.; Perumal, S.; Biswas, K. Ultrahigh Thermoelectric Figure of Merit and Enhanced Mechanical Stability of p-type $AgSb1-xZnxTe_2$. *ACS Energy Lett.* **2017**, *2*, 349–356. [CrossRef]
62. Du, B.; Li, H.; Xu, J.; Tang, X.; Uher, C. Enhanced Figure-of-Merit in Se-Doped p-Type $AgSbTe_2$ Thermoelectric Compound. *Chem. Mater.* **2010**, *22*, 5521–5527. [CrossRef]
63. Wang, H.; Li, J.-F.; Zou, M.; Sui, T. Synthesis and Transport Property of $AgSbTe_2$ as a Promising Thermoelectric Compound. *Appl. Phys. Lett.* **2008**, *93*, 202106. [CrossRef]
64. Gallo, C.F.; Chandrasekhar, B.S.; Sutter, P.H. Transport Properties of Bismuth Single Crystals. *J. Appl. Phys.* **1963**, *34*, 144–152. [CrossRef]
65. Xu, J.; Li, H.; Du, B.; Tang, X.; Zhang, Q.; Uher, C. High Thermoelectric Figure of Merit and Nanostructuring in Bulk $AgSbTe_2$. *J. Mater. Chem.* **2010**, *20*, 6138–6143. [CrossRef]
66. Wojciechowski, K.T.; Schmidt, M. Structural and Thermoelectric Properties of $AgSbTe-AgSbSe_2$ Pseudobinary System. *Phys. Rev. B* **2009**, *79*, 184202. [CrossRef]
67. Kosuga, A.; Uno, M.; Kurosaki, K.; Yamanaka, S. Thermoelectric properties of stoichiometric $Ag_{1-x}Pb_{18}SbTe_{20}$ ($x = 0, 0.1, 0.2$). *J. Alloys Compd.* **2005**, *391*, 288–291. [CrossRef]
68. Zhang, S.N.; Zhu, T.J.; Yang, S.H.; Yu, C.; Zhao, X.B. Phase Compositions, Nanoscale Microstructures and Thermoelectric Properties in $Ag_{2-y}SbyTe_{1+y}$ Alloys with Precipitated Sb_2Te_3 Plates. *Acta Mater.* **2010**, *58*, 4160–4169. [CrossRef]

Article

Pre-Ball-Milled Boron Nitride for the Preparation of Boron Nitride/Polyetherimide Nanocomposite Film with Enhanced Breakdown Strength and Mechanical Properties for Thermal Management

Ruiyi Li [1], Xiao Yang [1,2], Jian Li [1], Ding Liu [2,3], Lixin Zhang [1], Haisheng Chen [2,3], Xinghua Zheng [2,3,*] and Ting Zhang [1,2,3,4,*]

[1] Nanjing Institute of Future Energy System, Nanjing 211135, China
[2] Institute of Engineering Thermophysics, Chinese Academy of Sciences, Beijing 100190, China
[3] University of Chinese Academy of Sciences, Beijing 100049, China
[4] Innovation Academy for Light-Duty Gas Turbine, Chinese Academy of Sciences, Beijing 100190, China
* Correspondence: zhengxh@iet.cn (X.Z.); zhangting@iet.cn (T.Z.)

Abstract: Modern electronics not only require the thermal management ability of polymer packaging materials but also need anti-voltage and mechanical properties. Boron nitride nanosheets (BNNS), an ideal thermally conductive and high withstand voltage (800 kV/mm) filler, can meet application needs, but the complex and low-yield process limits their large-scale fabrication. Herein, in this work, we prepare sucrose-assisted ball-milled BN(SABM-BN)/polyetherimide (PEI) composite films by a casting-hot pressing method. SABM-BN, as a pre-ball-milled filler, contains BNNS and BN thick sheets. We mainly investigated the thermal conductivity (TC), breakdown strength, and mechanical properties of composites. After pre-ball milling, the in-plane TC of the composite film is reduced. It decreases from 2.69 to 2.31 W/mK for BN/PEI composite film at 30 wt% content; however, the through-plane TC of composites is improved, and the breakdown strength and tensile strength of the composite film reach the maximum of 54.6 kV/mm and 102.7 MPa at 5 wt% content, respectively. Moreover, the composite film is used as a flexible circuit substrate, and the working surface temperature is 20 °C, which is lower than that of pure PEI film. This study provides an effective strategy for polymer composites for electronic packaging.

Keywords: polymer composites; boron nitride nanosheets; thermal conductivity; breakdown strength; mechanical properties

1. Introduction

As electronic devices become increasingly miniaturized and intelligent, their energy density rises sharply and can be up to 300 W/cm^2 [1–3]. Local "hot spots" will have an irreversible impact on the life and stable use of electronic components. Because of their packaging and thermal management capabilities, thermally conductive polymer films are increasingly attracting attention, with the expectation of expanding their applications in flexible display screens [4], circuit substrates [5], and highly integrated insulated gate bipolar transistors (IGBT) [6].

Common insulating polymer films in electronic devices include polyimide (PI) [7], epoxy (EP) resin [8], and polyvinylidene fluoride (PVDF) [9]. However, the intrinsic thermal conductivity of this type of polymer is low, <0.2 W/mK [10]. It is often necessary to compound thermally conductive fillers to prepare thermal management materials (TMMs). Nevertheless, modern electronics also impose anti-voltage and mechanical performance requirements on TMMs to ensure low electrical failure rates and high reliability [11,12].

Boron nitride nanosheets (BNNS) have a two-dimensional hexagonal structure similar to graphene, high thermal conductivity (600 W/mK) [13–15], and ultra-high breakdown strength (800 kV/mm) [16], which are ideal choices for thermal management and anti-voltage polymer composites. Song et al. [17] incorporated 2 wt% BNNS into PVDF. The thermal conductivity was improved to 0.28 W/mK, and the breakdown strength increased from 366 kV/mm of pure film to 450 kV/mm. The common preparation process of BNNS includes ball milling [18], chemical exfoliation [19], and tip sonicate [20]. Note that these methods have so far been too complex, inefficient (~10 wt% yield), and very unsuitable for large-scale manufacturing. Still, the corresponding unexfoliated BN is more likely to form heat conduction networks than small-sized BNNS [21–23]. The size difference between bulk BN and BNNS may bring related synergies. Gu et al. [24] employed hybrid fillers of micrometer BN and nanometer BN with polyphenylene sulfide to hot-press into the highly thermal conductive composites. It was found that hybrid fillers were more beneficial in improving the thermal conductivity of composites compared to single-size fillers. These advantages are ignored, and the bulk BN is wasted by blindly pursuing BNNS. In addition, the functionalization of fillers, such as hydroxyl [25] or amino groups [26], should also be considered in the exfoliated process to increase the affinity with the polymer matrix and improve the mechanical properties [27–29].

Consequently, in this work, we use commercially available and cheap sucrose as grinding for exfoliating and modifying BN. Furthermore, polyetherimide (PEI), a mature and commercial PI, is selected as the matrix. The thermal conductivity, breakdown strength, and mechanical properties of composite films before and after pre-ball milling BN are emphatically compared. The film prepared from BN by sucrose-assisted ball milling is named SABM-BN/PEI composite film, and the untreated version is named BN/PEI composite film. As a result, the in-plane thermal conductivity of the SABM-BN/PEI composite film decreases to a certain extent compared with the BN/PEI composite film. However, the through-plane thermal conductivity of the SABM-BN/PEI composite film is improved. The breakdown strength of the SABM-BN/PEI composite film is higher than that of the BN/PEI composite film in the entire content range. For mechanical properties, the tensile strength of the composite film shows a similar trend to the breakdown strength. This improvement is attributed to the interface interaction. Verifying the thermal management capability of the SABM-BN/PEI composite film, the film surface uses screen printing to prepare a flexible circuit. The surface had a lower working temperature compared with commercial paper and PEI. Compared with previous work, this paper does not carry out subsequent separation after ball milling. The composite film prepared by combining the advantages of BNNS and BN has anti-voltage, excellent mechanical properties, and thermal management ability. The above shows that the prepared SABM-BN/PEI composite film has great potential to be used in electronics packaging and unique applications.

2. Materials and Methods

2.1. Materials

Polyetherimide powder (PEI, Ultem 1000) was purchased from SABIC Innovative Plastics Co., Ltd. (Shanghai, China). Boron nitride platelets (BN, ~20 µm) were supplied by Dandong Rijin Science and Technology Co., Ltd. (Dandong, China). N,N-dimethyl formamide (DMF) and sugar were provided by Shanghai Aladdin Biochemical Technology Co., Ltd. (Shanghai, China).

2.2. Preparation of Sugar-Assisted Ball-Milled BN Powder

In the dry ball milling process [30,31], BN (1 g), sugar (5 g), and zirconia balls (60 g) were loaded into a 500 mL grinding jar. The zirconia balls were divided into three kinds with diameters of 10 mm, 5 mm, and 1 mm in a weight ratio of 2:5:3. The mixture was ball milled at 500 rpm for 24 h using a planetary ball mill (YXQM-2L, MITR, China). Then D.I. water (200 mL) was poured into the mixture and stirred to speed up the dissolution of the sucrose. Finally, the suspension was vacuum filtered through an aqueous filter membrane (pore size: 0.22 µm) and washed several times to remove residual sucrose, followed by vacuum drying to obtain sugar-assisted ball-milled BN (SABM-BN). The SABM-BN contained the BN thick sheets (BNTS) and BN nanosheets (BNNS).

2.3. Preparation of PEI Composite Films

The preparation process of the BN/PEI and SABM-BN/PEI composite films is shown in Figure 1. First, BN or SABM-BN was ultrasonically dispersed in DMF (40 mL). Next, PEI powder (10 g) was added to the aforementioned DMF to dissolve. It was stirred at 80 °C for 4 h and sonicated to obtain a homogeneous composite solution. Then, the solution was cast on a clean glass plate. The glass plate was placed in an oven (160 °C) to remove the DMF. Subsequently, the film was peeled off the glass plate and shredded. The cut samples were placed between two layers of stainless-steel plates, and the mold release paper was Teflon fiber cloth. Finally, it was hot-pressed at 1 MPa, 270 °C for 10 min using a hot-pressing machine (6170B, BOLON, Dongguan, China) to obtain PEI composite films. Filler content was set at 0, 5, 10, 20, and 30 wt%.

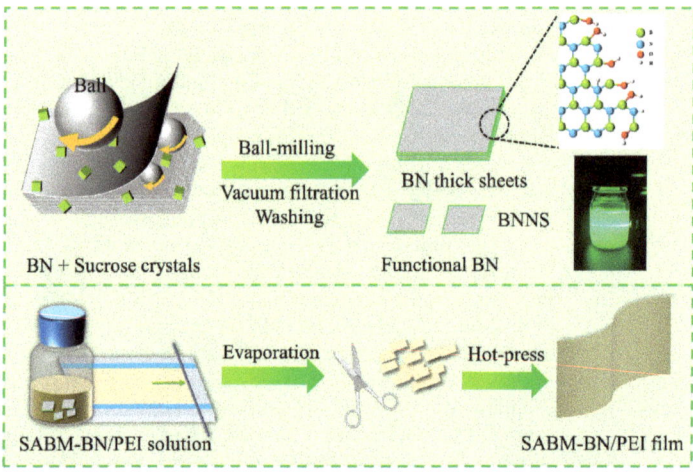

Figure 1. Schematic diagram illustrating the preparation of the SABM-BN and composite film.

2.4. Characterization

The Fourier transform infrared (FT-IR, Tensor 27, Bruker Nano Gmbh, Berlin, Germany) spectrum of the BN, SABM-BN, and sugar were collected using pressed KBr pellets. The filler and film cross-section morphologies were observed using a field emission scanning electronic microscope (FESEM, Regulus 8100, Hitachi, Tokyo, Japan). Before testing, the two samples were sprayed with a thin layer of gold for 90 s and 120 s. The differential scanning calorimetry (DSC, 204F1, Netzsch, Zelb, Germany) was performed to investigate the glass transition temperature (Tg). The films were heated to 300 °C at a 10 K/min heating rate under N_2 atmosphere. The mechanical properties of the composite films were measured on an electronic universal testing machine (105D-TS, Wance, Shenzhen, China) with a tensile rate of 1 mm/min. The breakdown strength was obtained using a withstand voltage tester (CD9917-AX, Changsheng, Nanjing, China) at a ramping voltage rate of 1 kV/s. A thermal constant analyzer (TPS2500s, Hot Disk AB, Västerås, Sweden) was used to determine the thermal conductivity of the composite film at room temperature. The heat resistance of the composite films was recorded using a thermal gravimetric analyzer (TG, 209F3, Netzsch, Zelb, Germany) from room temperature to 800 °C at N_2 atmosphere. To investigate the heat dissipation of the composite films in flexible electronic devices, first, silver circuits were screen-printed on the surface of the samples, followed dried. Next, a constant 12 V voltage was applied to the sample surface circuit, and the thermal distribution on the surface was simultaneously recorded using an infrared thermal imager (PS400, Guide, Wuhan, China).

3. Results and Discussion

Figure 1 shows the preparation process of the SABM-BN/PEI composite film. First, sucrose and BN are ball milled to obtain the SABM-BN, followed by casting and hot pressing to prepare the composite films. It is worth mentioning that, since the PEI solution easily absorbs moisture and causes irreversible phase transition during the casting process, it is necessary to further hot-press to obtain a dense composite film. As shown in Figure 2a, the peaks at 1375 cm^{-1} and 809 cm^{-1} on BN and SABM-BN correspond to the stretching and bending vibrations of B–N [32,33], respectively. It is found that some characteristic peaks of sucrose do not appear on the SABM-BN curve, which also indicates that the sucrose is completely removed during washing. In addition, on the SABM-BN curve, the new broad peak at 3431 cm^{-1} corresponds to the vibration of –OH [34]. For the products of saccharide-assisted ball-milled BN, extensive theoretical studies believe that H atoms in saccharides are preferentially combined with N atoms on BN to form N–H bonds [30,35–37], and the N–H bond on BN is easily hydrolyzed to generate NH_3 and –OH. For this reason, the slightly alkaline pH of the BN-sucrose suspension (Figure 2c) can better verify the sucrose modification. Functionalized BN will improve the adhesion between the filler and the polymer matrix.

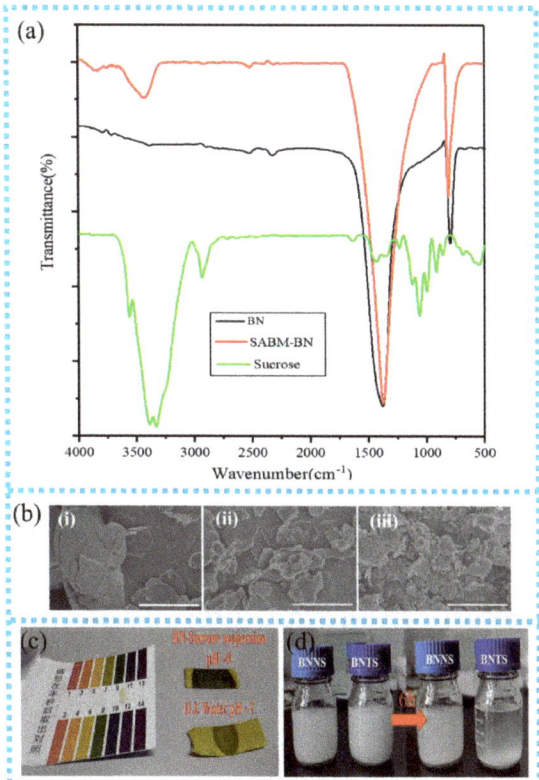

Figure 2. (**a**) FT-IR spectra of BN, sucrose, and SABM-BN. (**b**) SEM images of (**i**) BN, (**ii**) BNTS, and (**iii**) BNNS, respectively (scale bar: 10 μm). (**c**) Detection pH of BN-Sucrose suspension and D.I. water. (**d**) Stability of BNNS and BNTS (~1 mg/mL) aqueous solution for 6 h.

The SABM-BN contains exfoliating BNNS and larger size BN considering the ball milling efficiency. Therefore, the suspension is ultrasonically bathed (50 W) for 30 min before filtration and washing. Then the suspension is centrifuged at 2000 rpm for 30 min to separate the precipitate and supernatant, followed by filtration, washing, and drying to obtain BNTS and BNNS, respectively. The yields of BNTS and BNNS are 75.98% and 21.18%, respectively. However, the yield of BNNS is 10.24% without adding a grinding agent (sucrose). Although under high shear forces and collisions with the balls, the self-lubricating effect of BN sheets greatly reduces the exfoliation efficiency [38]. The addition of sucrose promotes the crushing of the sheets. The SEM images of BN, BNTS, and BNNS are shown in Figure 2b. The obtained BNNS has a Tyndall effect in an aqueous solution (~1 mg/mL, Figure 1). At the same concentration, BNNS has better dispersion stability than BNTS (Figure 2d) because larger size BNTS is more easily settled by gravity. All the above indicate that sucrose has a good effect on the exfoliation and modification of BN.

3.1. Thermal Properties Analysis

As mentioned above, the size of BN changed after ball milling. This will have an impact on the thermal conductivity of the composite film. Considering the anisotropy of film and sheet-like BN, the thermal conductivity of the composite film needs to be analyzed from both the horizontal and the vertical direction. Figure 3a shows the in-plane thermal conductivity ($\lambda_\parallel$) of the PEI and composite films. The $\lambda_\parallel$ of the 30 wt% BN/PEI composite film is 2.69 W/mK. The $\lambda_\parallel$ of the SABM-BN/PEI composite film is lower than that of the BN/PEI composite film. The $\lambda_\parallel$ of the 30 wt% SABM-BN/PEI composite film is 2.31 W/mK, which is ~11 times that of the pure film (0.205 W/mK). The through-plane thermal conductivity ($\lambda_\perp$) of the PEI and composite films (Figure 3b) is much lower than that of the in-plane, which is owing to the orientation of molecular chain arrangement [39] and the low $\lambda_\perp$ of BN (~10 W/mK) [32]. At the same time, it is found that $\lambda_\perp$ of the SABM-BN/PEI composite film is improved. At 30 wt% content, the $\lambda_\perp$ is 0.512 W/mK, higher than that of the BN/PEI composite film. We use the heat conduction model (Figure 3c) to explain. The large-sized BN tends to be arranged in parallel during the hot-pressing process. It is easier to form a thermal conduction network in the horizontal direction while the vertical direction is missing. However, after ball milling, the inconsistency of BN size destroys the continuity in the horizontal direction on the one hand. On the other hand, it gives the possibility to form a thermal network in the vertical direction.

Next, as shown in Figure 3d, the typical upper and lower tangent center is the glass transition temperature (Tg) of the polymer. The Tg of PEI is 215.6 °C. The Tg of the SABM-BN/PEI composite films is higher than that of PEI. Furthermore, it is found that the Tg of the pre-ball-milling BN/PEI composite film without sucrose does not change significantly. It may be because the surface –OH of SABM-BN will also strengthen the interaction with polymer molecular chains, hindering the movements of PEI molecular chains and thereby increasing the Tg [40,41].

Under high power density, local heat accumulation in electronic devices and expansion for unique application scenarios, such as aerospace, military [10], etc., will place requirements on the heat resistance of the substrate. As an ether group-containing PI, PEI itself has excellent heat resistance. The TG curves of the composite films are shown in Figure 3e, with the accelerated decomposition of the films at ~550 °C. The increasing residual rate with the increasing content indicates that the heat resistance of the composite films is further improved than that of the PEI film [42].

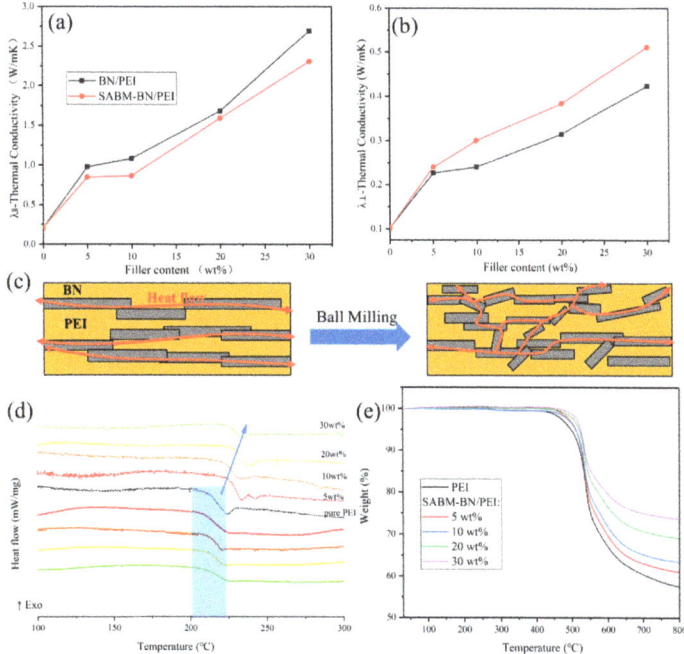

Figure 3. (**a**) In-plane and (**b**) through-plane thermal conductivity of PEI and the PEI composite films and (**c**) a heat conduction model. (**d**) DSC curves of PEI, the SABM-BN/PEI composite films (dashed line), and the pre-ball-milling BN/PEI composite film without sucrose (solid line). (**e**) TG curves of PEI and the SABM-BN/PEI composite films.

3.2. Application Assurance

Packaging substrates have excellent anti-voltage properties, which is the premise of ensuring the reliable operation of electronic devices. The anti-voltage properties of the composite films are tested by breakdown strength. The breakdown strength values of PEI and the PEI composite films are fitted by the Weibull distribution formula [43], as depicted in Equations (1) and (2).

$$P = 1 - exp\left[-\left(\frac{E_i}{E_0}\right)^\beta\right] \quad (1)$$

$$P = \frac{i - 0.5}{n + 0.25} \quad (2)$$

where P is the cumulative probability of electric failure, and E_i is the i-th breakdown strength after the measured values are arranged from small to large. n represents the number of electrical breakdown points of the sample, and here are eight data points. E_0 is the Weibull breakdown strength at $p = 63.2\%$ under the linear fit of Equation (1). β is the shape distribution parameter, representing the discrete situation of the data.

Figure 4a,b show the Weibull distribution plots and breakdown strengths of PEI and the PEI composite films. As shown in Figure 4b, (i) the breakdown strength of the SABM-BN/PEI composite films is generally higher than that of the BN/PEI composite films, and the maximum value (54.6 kV/mm) appears when the content is 5 wt%. This is mainly owing to the contribution of BNNS. BNNS has an ultra-high theoretical withstand voltage performance (800 kV/mm). At the same time, after ball milling, the size reduction of BN will make the development of the electric tree during the breakdown process more tortuous [44]. Thus, the breakdown strength of the composite film is improved. (ii) With the increase of filler content, the breakdown strength of the films decreases from 49.4 kV/mm of PEI to 36.2 and 38.5 kV/mm. Since the increase of filler content will bring more defects, the generation of conductive paths will result [45]. Although the addition of high-content fillers reduces the breakdown strength of the composite film, it still meets the anti-voltage requirements in electronic devices (>5 kV/mm) [17].

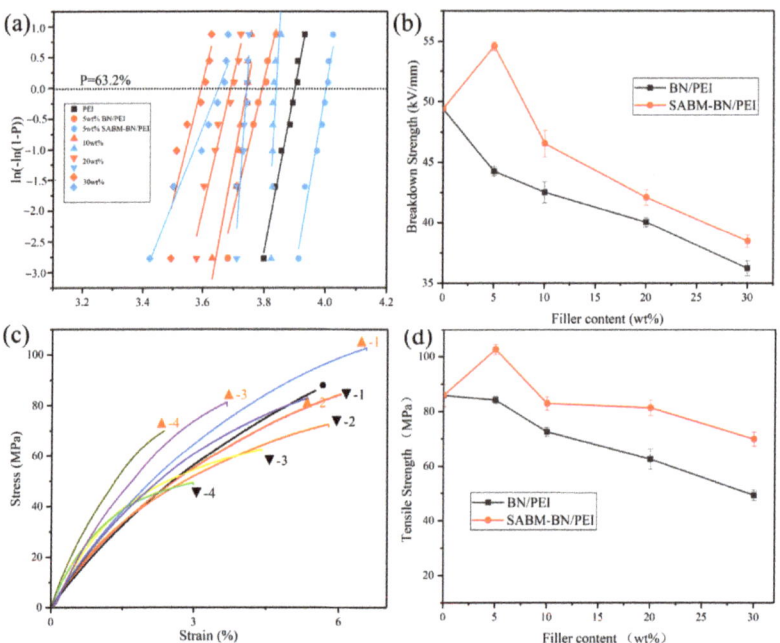

Figure 4. (**a**) Weibull distribution plots and (**b**) breakdown strength of PEI and the PEI composite films. (**c**) The stress-strain curves and (**d**) tensile strength of PEI and the PEI composite films. Black circle: PEI. Downward-facing black triangle 1–4: 5~30 wt% BN/PEI composite films. Upward-facing orange triangle 1–4: 5~30 wt% SABM-BN/PEI composite films.

Figure 4c shows stress-strain curves of PEI and the PEI composite films. The SABM-BN/PEI composite film curve is higher than that of the BN/PEI composite film at the same content. The tensile strength of the composite films is similar to the breakdown strength change (Figure 4d). The tensile strength of the PEI film is 85.87 MPa. The mechanical properties of the BN/PEI composite films decrease with the increasing content. At 30 wt% content, the tensile strength of the composite film is 49.38 MPa. The SABM-BN/PEI composite films have better mechanical properties than the BN/PEI composite films, and a typical "rise-fall" process occurs: the tensile strength of the 5 wt% SABM-BN/PEI composite film is 102.6 MPa, and at 30 wt% content, the tensile strength is 69.83 MPa. The improvement in tensile strength is attributed to the hydroxyl-rich surface of BN after ball milling, which promotes the compatibility of fillers with a polymer matrix and reduces stress concentration points [46].

3.3. Morphological Distribution

Figure 5 shows the cross-section morphologies of PEI and the SABM-BN/PEI composite films. The cross-section of pure PEI film is denser than that of composite films. It is found that with the increase of filler content, the filler arrangement tends to be more horizontal. At 5 wt% content, the distribution of fillers has no obvious regularity, but at 20 wt% content, the distribution of fillers begins to tend to be horizontal. This is because the mutual volume between the fillers becomes more significant at high filler content so that it is distributed in parallel in the film under the external force of hot pressing [47]. It effectively explains the reason for the sudden increase in the in-plane thermal conductivity of the composite film at 20 wt% content; the heat conduction network in the horizontal direction has been formed. As shown in Figure 5f, the distribution of fillers of different sizes in the films is observed at high magnifications.

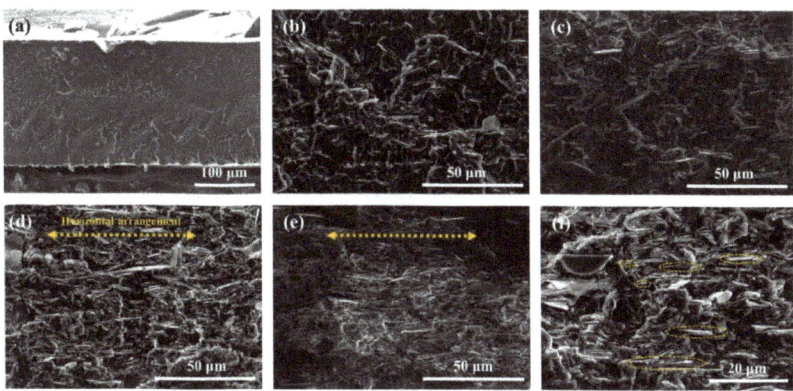

Figure 5. Cross-section SEM images of (**a**) PEI, (**b**) the 5 wt%, (**c**) 10 wt%, (**d**) 20 wt%, and (**e**) 30 wt% SABM-BN/PEI composite film. (**f**) Filler distribution at high magnification for the 20 wt% SABM-BN/PEI composite film.

3.4. Thermal Management

To evaluate the thermal management ability of the prepared composite film as a flexible substrate, here we use screen printing to print circuits on the surface of the composite films (30 wt% SABM-BN/PEI). Paper and PEI film serve as controls. As shown in Figure 6a, the substrate size is tailored to 4 cm × 6.5 cm, the flexible circuit integrates with the PEI composite film that can be bent and has a certain flexibility, and then a constant voltage of 12 V is applied across the circuit. Owing to the existence of Joule heat, the film surface heats up immediately, and the surface heat distribution is shown in Figure 6c. It is found that the PEI film has a fast temperature rise rate, which is stable at around ~60 °C, followed by the paper, which is stable at ~45 °C. Importantly, the surface temperature of the PEI composite film was ~20 °C lower than that of the pure film, indicating that the "hot spot" temperature can be effectively reduced. Such a composite film substrate has excellent thermal management capability and can be used in flexible electronic devices with high energy density.

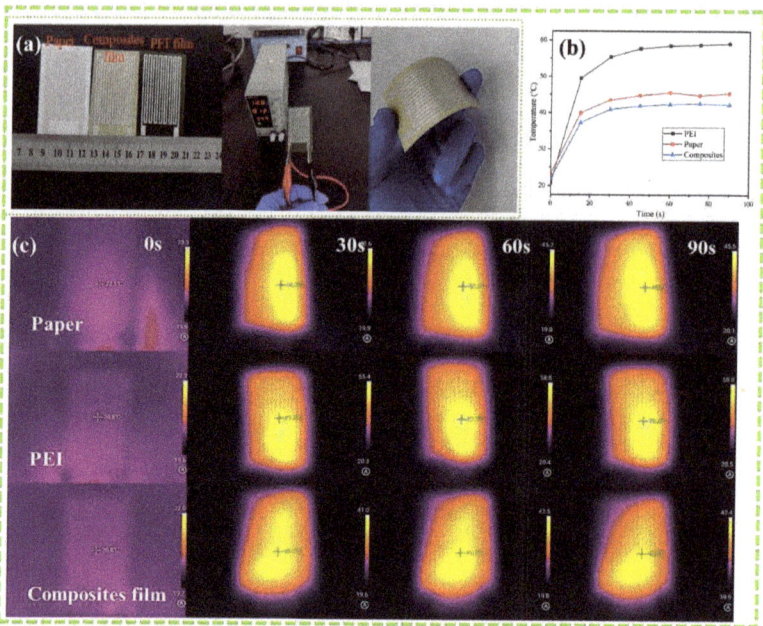

Figure 6. (a) From left to right: optical photographs of flexible circuits integrated into paper, PEI, and composite film; applying voltage; and bending the flexible substrate. (b) Surface temperature versus time. (c) Infrared thermal images.

4. Conclusions

In summary, commercially available and cheap sucrose was chosen to exfoliate and modify BN simultaneously. The yield of BNNS in SABM-BN is ~21%, and SABM-BN has –OH groups. The size changes after ball milling and results in opposite differences in $\lambda_\parallel$ and $\lambda_\perp$ of the SABM-BN/PEI composite film and the BN/PEI composite film. The $\lambda_\parallel$ of the SABM-BN/PEI composite film decreases to a certain extent compared with the BN/PEI composite film. At 30 wt% content, the BN/PEI composite film is 2.69 W/mK, while the SABM-BN/PEI composite film is 2.31 W/mK. However, the through-plane thermal conductivity of the SABM-BN/PEI composite film is improved. After ball-milling, the thermal conductive network reduces in the horizontal direction but becomes feasible in the vertical direction. Owing to the presence of BNNS and the modification of sucrose, the composite films' breakdown strength and mechanical properties are improved. The maximum reaches 54.6 kV/mm and 102.7 MPa at 5 wt% content. Furthermore, as a flexible circuit substrate, the composite film has excellent thermal management capability. The working surface temperature is 20 °C lower than that of pure PEI film. We believe that this study provides an effective strategy for high-performance polymer composites for electronic packaging.

Author Contributions: Conceptualization, J.L., D.L. and L.Z.; Data curation, X.Y.; Funding acquisition, T.Z.; Investigation, R.L.; Methodology, R.L.; Project administration, T.Z.; Software, J.L., D.L. and L.Z.; Supervision, X.Z.; Validation, X.Y., J.L., D.L. and L.Z.; Visualization, H.C.; Writing—review & editing, R.L. and T.Z. All authors have read and agreed to the published version of the manuscript.

Funding: This work was funded by the National Natural Science Foundation of China (NO.52172249 and NO.51976215), the Special Funding of Carbon Peak and Neutrality Science and Technology Innovation Project of Jiangsu Province (BE2022011), the Funding of Nanjing Institute of Future Energy System (KCW-12), the Scientific Instrument Developing Project of the Chinese Academy of

Sciences (YJKYYQ20200017), and the Funding of Innovation Academy for Light-duty Gas Turbine, Chinese Academy of Sciences (CXYJJ21-ZD-02).

Institutional Review Board Statement: Not applicable.

Informed Consent Statement: Not applicable.

Data Availability Statement: Not applicable.

Acknowledgments: We gratefully acknowledge Pei Huang from the College of Chemical Engineering of Nanjing Tech University for FT-IR, DSC, and mechanical testing.

Conflicts of Interest: The authors declare no conflict of interest.

References

1. Zhang, X.-D.; Yang, X.-H.; Zhou, Y.-X.; Rao, W.; Gao, J.-Y.; Ding, Y.-J.; Shu, Q.-Q.; Liu, J. Experimental investigation of galinstan based minichannel cooling for high heat flux and large heat power thermal management. *Energy Convers. Manag.* **2019**, *185*, 248–258. [CrossRef]
2. Lee, D.; Lee, H.; Song, T.; Paik, U. Toward High Rate Performance Solid–State Batteries. *Adv. Energy Mater.* **2022**, *12*, 2200948. [CrossRef]
3. Li, R.; Yang, X.; Li, J.; Shen, Y.; Zhang, L.; Lu, R.; Wang, C.; Zheng, X.; Chen, H.; Zhang, T. Review on polymer composites with high thermal conductivity and low dielectric properties for electronic packaging. *Mater. Today Phys.* **2022**, *22*, 100594. [CrossRef]
4. Xing, L.; Luscombe, C.K. Advances in applying C–H functionalization and naturally sourced building blocks in organic semiconductor synthesis. *J. Mater. Chem. C* **2021**, *9*, 16391–16409. [CrossRef]
5. Xiang, J.; Zhou, G.; Hong, Y.; He, W.; Wang, S.; Chen, Y.; Wang, C.; Tang, Y.; Sun, Y.; Zhu, Y. Direct additive copper plating on polyimide surface with silver ammonia via plasma modification. *Appl. Surf. Sci.* **2022**, *587*, 152848. [CrossRef]
6. Wang, B.; Wang, L.; Mu, W.; Qin, M.; Yang, F.; Liu, J.; Tomoyuki, Y.; Tatsuhiko, F. Thermal Performances and Annual Damages Comparison of MMC Using Reverse Conducting IGBT and Conventional IGBT Module. *IEEE Trans. Power Electron.* **2021**, *36*, 9806–9825. [CrossRef]
7. Wu, X.; Li, H.; Cheng, K.; Qiu, H.; Yang, J. Modified graphene/polyimide composite films with strongly enhanced thermal conductivity. *Nanoscale* **2019**, *11*, 8219–8225. [CrossRef]
8. Kim, K.; Kim, J. Exfoliated boron nitride nanosheet/MWCNT hybrid composite for thermal conductive material via epoxy wetting. *Compos. Part B Eng.* **2018**, *140*, 9–15. [CrossRef]
9. Wu, L.; Luo, N.; Xie, Z.; Liu, Y.; Chen, F.; Fu, Q. Improved breakdown strength of Poly(vinylidene fluoride)-based composites by using all ball-milled hexagonal boron nitride sheets without centrifugation. *Compos. Sci. Technol.* **2020**, *190*, 108046. [CrossRef]
10. Guo, Y.; Ruan, K.; Shi, X.; Yang, X.; Gu, J. Factors affecting thermal conductivities of the polymers and polymer composites: A review. *Compos. Sci. Technol.* **2020**, *193*, 108134. [CrossRef]
11. Wu, M.; Yang, L.; Zhou, Y.; Jiang, J.; Zhang, L.; Rao, T.; Yang, P.; Liu, B.; Liao, W. BaTiO$_3$-assisted exfoliation of boron nitride nanosheets for high-temperature energy storage dielectrics and thermal management. *Chem. Eng. J.* **2022**, *427*, 131860. [CrossRef]
12. Ren, L.; Zeng, X.; Sun, R.; Xu, J.-B.; Wong, C.-P. Spray-assisted assembled spherical boron nitride as fillers for polymers with enhanced thermally conductivity. *Chem. Eng. J.* **2019**, *370*, 166–175. [CrossRef]
13. Xie, Z.; Wu, K.; Liu, D.; Zhang, Q.; Fu, Q. One-step alkyl-modification on boron nitride nanosheets for polypropylene nanocomposites with enhanced thermal conductivity and ultra-low dielectric loss. *Compos. Sci. Technol.* **2021**, *208*, 108756. [CrossRef]
14. Guerra, V.; Wan, C.; McNally, T. Thermal conductivity of 2D nano-structured boron nitride (BN) and its composites with polymers. *Prog. Mater. Sci.* **2019**, *100*, 170–186. [CrossRef]
15. Bai, X.; Zhang, C.; Zeng, X.; Ren, L.; Sun, R.; Xu, J. Recent progress in thermally conductive polymer/boron nitride composites by constructing three-dimensional networks. *Compos. Commun.* **2021**, *24*, 100650. [CrossRef]
16. Ayoob, R.; Alhabill, F.; Andritsch, T.; Vaughan, A. Enhanced dielectric properties of polyethylene/hexagonal boron nitride nanocomposites. *J. Mater. Sci.* **2017**, *53*, 3427–3442. [CrossRef]
17. Song, Q.; Zhu, W.; Deng, Y.; Zhu, M.; Zhang, Q. Synergetic optimization of thermal conductivity and breakdown strength of boron nitride/poly (vinylidene fluoride) composite film with sandwich intercalated structure for heat management in flexible electronics. *Compos. Part A Appl. Sci. Manuf.* **2020**, *135*, 105933. [CrossRef]
18. Han, G.; Zhao, X.; Feng, Y.; Ma, J.; Zhou, K.; Shi, Y.; Liu, C.; Xie, X. Highly flame-retardant epoxy-based thermal conductive composites with functionalized boron nitride nanosheets exfoliated by one-step ball milling. *Chem. Eng. J.* **2021**, *407*, 127099. [CrossRef]
19. Zhang, Y.; Fan, Y.; Kamran, U.; Park, S.-J. Improved thermal conductivity and mechanical property of mercapto group-activated boron nitride/elastomer composites for thermal management. *Compos. Part A Appl. Sci. Manuf.* **2022**, *156*, 106869. [CrossRef]
20. Yu, J.; Mo, H.; Jiang, P. Polymer/boron nitride nanosheet composite with high thermal conductivity and sufficient dielectric strength. *Polym. Adv. Technol.* **2015**, *26*, 514–520. [CrossRef]
21. Pan, C.; Zhang, J.; Kou, K.; Zhang, Y.; Wu, G. Investigation of the through-plane thermal conductivity of polymer composites with in-plane oriented hexagonal boron nitride. *Int. J. Heat Mass Transf.* **2018**, *120*, 1–8. [CrossRef]

22. Yung, K.C.; Liem, H. Enhanced thermal conductivity of boron nitride epoxy-matrix composite through multi-modal particle size mixing. *J. Appl. Polym. Sci.* **2007**, *106*, 3587–3591. [CrossRef]
23. Ishida, H.; Rimdusit, S. Very high thermal conductivity obtained by boron nitride-filled polybenzoxazine. *Thermochim. Acta* **1998**, *320*, 177–186. [CrossRef]
24. Gu, J.; Guo, Y.; Yang, X.; Liang, C.; Geng, W.; Tang, L.; Li, N.; Zhang, Q. Synergistic improvement of thermal conductivities of polyphenylene sulfide composites filled with boron nitride hybrid fillers. *Compos. Part A Appl. Sci. Manuf.* **2017**, *95*, 267–273. [CrossRef]
25. Sato, K.; Horibe, H.; Shirai, T.; Hotta, Y.; Nakano, H.; Nagai, H.; Mitsuishi, K.; Watari, K. Thermally conductive composite films of hexagonal boron nitride and polyimide with affinity-enhanced interfaces. *J. Mater. Chem.* **2010**, *20*, 2749–2752. [CrossRef]
26. Pan, C.; Kou, K.; Jia, Q.; Zhang, Y.; Wu, G.; Ji, T. Improved thermal conductivity and dielectric properties of hBN/PTFE composites via surface treatment by silane coupling agent. *Compos. Part B Eng.* **2017**, *111*, 83–90. [CrossRef]
27. Zhou, S.; Xu, T.; Jiang, F.; Song, N.; Shi, L.; Ding, P. High thermal conductivity property of polyamide-imide/boron nitride composite films by doping boron nitride quantum dots. *J. Mater. Chem. C* **2019**, *7*, 13896–13903. [CrossRef]
28. Korycki, A.; Chabert, F.; Merian, T.; Nassiet, V. Probing Wettability Alteration of the Boron Nitride Surface through Rheometry. *Langmuir* **2019**, *35*, 128–140. [CrossRef]
29. Agrawal, A.; Chandrakar, S. Influence of particulate surface treatment on physical, mechanical, thermal, and dielectric behavior of epoxy/hexagonal boron nitride composites. *Polym. Compos.* **2020**, *41*, 1574–1583. [CrossRef]
30. Chen, S.; Xu, R.; Liu, J.; Zou, X.; Qiu, L.; Kang, F.; Liu, B.; Cheng, H.M. Simultaneous Production and Functionalization of Boron Nitride Nanosheets by Sugar-Assisted Mechanochemical Exfoliation. *Adv Mater* **2019**, *31*, e1804810. [CrossRef]
31. Balasubramanyan, S.; Sasidharan, S.; Poovathinthodiyil, R.; Ramakrishnan, R.M.; Narayanan, B.N. Sucrose-mediated mechanical exfoliation of graphite: A green method for the large scale production of graphene and its application in catalytic reduction of 4-nitrophenol. *New J. Chem.* **2017**, *41*, 11969–11978. [CrossRef]
32. Ding, D.; Shang, Z.; Zhang, X.; Lei, X.; Liu, Z.; Zhang, Q.; Chen, Y. Greatly enhanced thermal conductivity of polyimide composites by polydopamine modification and the 2D-aligned structure. *Ceram. Int.* **2020**, *46*, 28363–28372. [CrossRef]
33. Boukheit, N.; Chabert, F.; Otazaghine, B.; Taguet, A. h-BN Modification Using Several Hydroxylation and Grafting Methods and Their Incorporation into a PMMA/PA6 Polymer Blend. *Nanomaterials* **2022**, *12*, 2735. [CrossRef] [PubMed]
34. Ge, M.; Zhang, J.; Zhao, C.; Lu, C.; Du, G. Effect of hexagonal boron nitride on the thermal and dielectric properties of polyphenylene ether resin for high-frequency copper clad laminates. *Mater. Des.* **2019**, *182*, 108028. [CrossRef]
35. Wang, Z.-G.; Wei, X.; Bai, M.-H.; Lei, J.; Xu, L.; Huang, H.-D.; Du, J.; Dai, K.; Xu, J.-Z.; Li, Z.-M. Green Production of Covalently Functionalized Boron Nitride Nanosheets via Saccharide-Assisted Mechanochemical Exfoliation. *ACS Sustain. Chem. Eng.* **2021**, *9*, 11155–11162. [CrossRef]
36. Li, Y.; Huang, T.; Chen, M.; Wu, L. Simultaneous exfoliation and functionalization of large-sized boron nitride nanosheets for enhanced thermal conductivity of polymer composite film. *Chem. Eng. J.* **2022**, *442*, 136237. [CrossRef]
37. Ramos, K.J.; Bahr, D.F. Mechanical behavior assessment of sucrose using nanoindentation. *J. Mater. Res.* **2011**, *22*, 2037–2045. [CrossRef]
38. Lei, W.; Mochalin, V.N.; Liu, D.; Qin, S.; Gogotsi, Y.; Chen, Y. Boron nitride colloidal solutions, ultralight aerogels and freestanding membranes through one-step exfoliation and functionalization. *Nat. Commun.* **2015**, *6*, 8849. [CrossRef]
39. Tanimoto, M.; Yamagata, T.; Miyata, K.; Ando, S. Anisotropic thermal diffusivity of hexagonal boron nitride-filled polyimide films: Effects of filler particle size, aggregation, orientation, and polymer chain rigidity. *ACS Appl. Mater. Interfaces* **2013**, *5*, 4374–4382. [CrossRef]
40. Bozkurt, Y.E.; Yıldız, A.; Türkarslan, Ö.; Şaşal, F.N.; Cebeci, H. Thermally conductive h-BN reinforced PEI composites: The role of processing conditions on dispersion states. *Mater. Today Commun.* **2021**, *29*, 102854. [CrossRef]
41. Gu, J.; Lv, Z.; Wu, Y.; Guo, Y.; Tian, L.; Qiu, H.; Li, W.; Zhang, Q. Dielectric thermally conductive boron nitride/polyimide composites with outstanding thermal stabilities via in -situ polymerization-electrospinning-hot press method. *Compos. Part A Appl. Sci. Manuf.* **2017**, *94*, 209–216. [CrossRef]
42. Madakbaş, S.; Çakmakçı, E.; Kahraman, M.V. Preparation and thermal properties of polyacrylonitrile/hexagonal boron nitride composites. *Thermochim. Acta* **2013**, *552*, 1–4. [CrossRef]
43. Zhuo, L.; Chen, S.; Xie, F.; Qin, P.; Lu, Z. Toward high thermal conductive aramid nanofiber papers: Incorporating hexagonal boron nitride bridged by silver nanoparticles. *Polym. Compos.* **2021**, *42*, 1773–1781. [CrossRef]
44. Peng, X.; Liu, X.; Qu, P.; Yang, B. Enhanced breakdown strength and energy density of PVDF composites by introducing boron nitride nanosheets. *J. Mater. Sci. Mater. Electron.* **2018**, *29*, 16799–16804. [CrossRef]
45. Luo, W.; Zeng, J.; Chen, Y.; Dai, W.; Yao, Y.; Luo, B.; Zhang, F.; Wang, T. Surface modification of h-BN and preparation of h-BN/PEI thermally conductive flexible films. *Polym. Compos.* **2022**, *43*, 3846–3857. [CrossRef]
46. Liu, X.; Ji, T.; Li, N.; Liu, Y.; Yin, J.; Su, B.; Zhao, J.; Li, Y.; Mo, G.; Wu, Z. Preparation of polyimide composites reinforced with oxygen doped boron nitride nano-sheet as multifunctional materials. *Mater. Des.* **2019**, *180*, 107963. [CrossRef]
47. Zhang, T.; Sun, J.; Ren, L.; Yao, Y.; Wang, M.; Zeng, X.; Sun, R.; Xu, J.-B.; Wong, C.-P. Nacre-inspired polymer composites with high thermal conductivity and enhanced mechanical strength. *Compos. Part A Appl. Sci. Manuf.* **2019**, *121*, 92–99. [CrossRef]

Article

High Thermoelectric Power Generation by SWCNT/PPy Core Shell Nanocomposites

M. Almasoudi [1,2], Numan Salah [3,4,*], Ahmed Alshahrie [1,4], Abdu Saeed [1], Mutabe Aljaghtham [5], M. Sh. Zoromba [6,7], M. H. Abdel-Aziz [6,8] and Kunihito Koumoto [4,9]

[1] Department of Physics, Faculty of Science, King Abdulaziz University, Jeddah 21589, Saudi Arabia; malmasoudi0020@stu.kau.edu.sa (M.A.); aalshahri@kau.edu.sa (A.A.); abdusaeed79@hotmail.com (A.S.)
[2] Department of Physics, Al-Qunfudah University College, Umm Al-Qura University, Makkah 21955, Saudi Arabia
[3] K. A. CARE Energy Research and Innovation Center, King Abdulaziz University, Jeddah 21589, Saudi Arabia
[4] Center of Nanotechnology, King Abdulaziz University, Jeddah 21589, Saudi Arabia; g44233a@cc.nagoya-u.ac.jp
[5] Department of Mechanical Engineering, College of Engineering, Prince Sattam bin Abdulaziz University, Al Kharj 16273, Saudi Arabia; m.aljaghtham@psau.edu.sa
[6] Department of Chemical and Materials Engineering, King Abdulaziz University, Rabigh 21911, Saudi Arabia; mzoromba@kau.edu.sa (M.S.Z.); mhmossa@kau.edu.sa (M.H.A.-A.)
[7] Department of Chemistry, Faculty of Science, Port Said University, Port-Said 42521, Egypt
[8] Department of Chemical Engineering, Faculty of Engineering, Alexandria University, Alexandria 5424041, Egypt
[9] Nagoya Industrial Science Research Institute, Nagoya 464-0819, Japan
* Correspondence: nsalah@kau.edu.sa or alnumany@yahoo.com

Abstract: Polypyrrole (PPy) is a conducting polymer with attractive thermoelectric (TE) properties. It is simple to fabricate and modify its morphology for enhanced electrical conductivity. However, such improvement is still limited to considerably enhancing TE performance. In this case, a single-wall carbon nanotube (SWCNT), which has ultrathin diameters and exhibits semi-metallic electrical conductivity, might be a proper candidate to be combined with PPy as a core shell one-dimensional (1D) nanocomposite for higher TE power generation. In this work, core shell nanocomposites based on SWCNT/PPy were fabricated. Various amounts of pyrrole (Py), which are monomer sources for PPy, were coated on SWCNT, along with methyl orange (MO) as a surfactant and ferric chloride as an initiator. The optimum value of Py for maximum TE performance was determined. The results showed that the SWCNT acted as a core template to direct the self-assembly of PPy and also to further enhance TE performance. The TE power factor, PF, and figure of merit, zT, values of the pure PPy were initially recorded as ~1 $\mu W/mK^2$ and 0.0011, respectively. These values were greatly increased to 360 $\mu W/mK^2$ and 0.09 for the optimized core shell nanocomposite sample. The TE power generation characteristics of the fabricated single-leg module of the optimized sample were also investigated and confirmed these findings. This enhancement was attributed to the uniform coating and good interaction between PPy polymer chains and walls of the SWCNT through π–π stacking. The significant enhancement in the TE performance of SWCNT/PPy nanocomposite is found to be superior compared to those reported in similar composites, which indicates that this nanocomposite is a suitable and scalable TE material for TE power generation.

Keywords: thermoelectric materials; conducting polymers; polypyrrole; single-wall carbon nanotubes; core shell nanocomposites

1. Introduction

In recent years, there has been rapid development in organic thermoelectric materials, especially conducting polymers and their composites due to several advantages over inorganic thermoelectric materials. These conducting polymers exhibit light weight, lower cost, reasonable thermal conductivity, easy fabrication process, and excellent flexibility. For instance, the thermoelectric performance of organic polymers can be significantly improved by controlling the combination of carbon nanotubes (CNTs) or graphene nanosheets with such polymers [1–3]. So far, the most common conducting polymers that have been investigated as thermoelectric materials are polyaniline (PANI) [4], polypyrrole (PPy) [5], and poly (3,4-ethylene dioxythiophene) (PEDOT) [6–8]. However, the TE performance was limited because of the low electrical conductivity of the used carbon nanostructures or due to the low interaction between the conducting polymer and carbon material [9,10]. Tuning the morphology of such polymers was also performed in [6,11,12]; however, no significant improvement in the TE performance was reported.

The conducted work on PPy-based composites as a TE material is less compared to that reported on PEDOT and PANI composites, but recent studies showed that when PPy is doped with appropriate dopants or included in a proper composite material, it shows good mechanical properties, high electrical conductivity, and low thermal conductivity, which therefore enhance the TE performance [13]. The nanostructure form of PPy was also reported to have a considerable TE performance [9,14–16], especially the PPy nanotube, which showed remarkable improvement in TE performance [13]. Multiwall carbon nanotubes (MWCNT) or single-wall carbon nanotubes (SWCNT) were also used to enhance the TE properties of PPy [10,17–19]. The effect of other carbon nanostructures such as graphene nanosheets [20] or reduced graphene oxide (rGO) [21] was also investigated to obtain the TE performance of the PPy. However, it is still necessary to explore other approaches or suitable precursors that can be developed, such as a one-dimensional (1D) core shell structure of CNTs/PPy, with a smooth coating. This coating might facilitate charge transport and thus increases the electrical conductivity. It also can generate extra energy filtering sites at the interfaces between the PPy and CNTs.

The recent work reported in our lab on PPy [22,23] and PPy with carbon nanotubes [20,22] was focused on enhancing the TE performance of this polymer, but these efforts were focused either on the effect of surfactant type [22] or the surfactant [23] and carbon nanotubes [18] concentrations on the TE performance of the PPy. Although these studies demonstrate the capability of this polymer as TE, the observed TE performance is still low for real application as TE materials. One of the most important factors to enhance the TE performance of PPy is the selection of highly conducting carbon nanotubes such as single-wall carbon nanotubes, which have not been well addressed. It is understood that the electrical conductivity of carbon nanotubes can vary from semiconductors to metallic. This TE property depends on several factors such as diameters and chirality, and even the SWCNT, which is considered the best interim of their electrical conductivity, can vary from low semi-conductive to metallics [24]. In this case, coating PPy with highly conducting SWCNT along with an electrical conductivity value larger than 50,000 S/m (in a pressed compact form) might be quite important to enhance the TE performance. Moreover, selecting the proper surfactant and oxidant are also important for developing a 1D core shell structure with a smooth coating.

In this work, PPy was synthesized in the presence of methyl orange (MO) as a surfactant to regulate its shape as a blank thermoelectric polymer. Subsequently, highly conducting SWCNTs along with various concentrations of PPy were produced in the form of 1D core shell nanocomposite structures (SWCNT/Ppy) using an in situ polymerization method. The manufactured SWCNT/PPy nanocomposites were characterized by several well-known techniques such as SEM, TEM, Raman, FTIR, and XPS spectroscopy to analyze the TE performance. Moreover, the power generation characteristics of a single-leg module of the optimized SWCNT/PPy core shell nanocomposite were numerically and experimentally quantified. Additionally, the TE performance of the present SWCNT/PPy nanocomposite is found to be superior compared to previous studies reported in the literature.

2. Materials and Methods

2.1. Materials

Highly pure pyrrole monomer (Py), ethanol, anhydrous ferric chloride ($FeCl_3 \cdot 6H_2O$), and methyl orange (MO) were purchased from Sigma-Aldrich, Steinheim, Germany. All reagents were of analytical grade (99.99%) and used as received without further purification. Single-wall carbon nanotubes (SWCNTs) of high electrical conductivity (>1000 S/cm in their pellet form) were purchased from Ad-Nano Technologies (Karnataka, India).

2.2. Synthesis of SWCNT/PPy Nanocomposites

Pure PPy was prepared by dissolving 750 mg of MO and 1 mL of Py monomer in 50 mL of absolute ethanol, and the resulting solution was diluted by DI water up to 200 mL using a magnetic stirrer (800 rpm) at room temperature (RT) for 20 min. In another beaker, 2340 mg of $FeCl_3 \cdot 6H_2O$ as an oxidant agent was dissolved in 200 mL of DI water, which was then added dropwise to the Py solution. The mixture was magnetically stirred for 48 h. The resulting black precipitation was filtered and washed several times successively with DI water and absolute ethanol to remove the surfactant and unreacted species. Then, the resulting product was dried at 60 °C for 24 h. The SWCNT/PPy nanocomposites were prepared as follows. A typical amount of SWCNT (300 mg) was dispersed in 50 mL of ethanol; then, this solution was diluted by 150 mL of DI water and sonicated for 3 h. The desired amounts of MO and Py monomer were added to the previous solution under continuous stirring for 30 min. Then, under vigorous stirring, the corresponding amount of $FeCl_3 \cdot 6H_2O$ in 200 mL of DI water was added slowly dropwise into the above solution to initiate the polymerization reaction and the mixture was magnetically stirred for two days. The product was filtered, washed, and dried exactly as in the case of pure PPy. Table 1 shows SWCNT amounts and the chemical materials used for this coating.

Table 1. Raw materials used for coating the SWCNT with PPy at a different layer thickness.

Samples	SWCNT (mg)	MO (mg)	Py (mL) (In 200 mL DI)	$FeCl_3$ (mg) (In 200 mL DI)
PW1		37.5	0.05	117
PW2		75	0.1	234
PW3	300 mg	375	0.5	1170
PW4		750	1	2340
PW5		1500	2	4680

2.3. Characterizations

The morphology and microstructure of the pristine SWCNT, the neat PPy, and SWCNT/PPy nanocomposites were investigated using scanning electron microscopy (SEM) (JSM-7500F, JEOL, Tokyo, Japan) and transmission electron microscopy (TEM) (JEM 2100F, JEOL). Raman spectra were acquired using micro-Raman spectroscopy (Thermo Fisher Scientific, Waltham, MA, USA), whereas the FTIR spectra of the samples were derived by attenuated total reflection–Fourier transform infrared (ATR-FTIR) spectrometry (Thermo Fisher Scientific, Waltham, MA, USA). X-ray diffraction (XRD) studies were carried out using an Ultima-IV X-ray diffractometer (Rigaku, Japan) equipped with Cu Kα radiation (k = 1.5406 Å), while the surface composition changes of the synthesized samples were examined using X-ray photoelectron spectroscopy (XPS, PHI 5000 VersaProbe, Japan). To investigate the thermoelectrical properties, the pellets of the pure and SWCNT/PPy nanocomposites were prepared under a pressure of 15 tons utilizing a manual hydraulic press. The prepared pellets, which have dimensions of 13 mm diameter and 1.5–2.0 mm thickness, were annealed at 370 K for one hour in an oven under a vacuum atmosphere, and then their thermoelectric properties were measured.

The electrical resistivity and Seebeck coefficient measurements from 300 to 370 K were accomplished using the LSR-3 (Linseis GmbH) in the He atmosphere. The heating rate was fixed at 5 K/min and the temperature gradient between the cold and hot sides was set at

50 K. Thermal conductivity of the fabricated pellets was then determined using the laser flash method in LFA-1000 (Linseis, Selb, Germany). The measurements were performed perpendicular to the surface of the pressed pellet in a vacuum atmosphere and the pellet sides' heating rate was set at 10 K/min. The charge carrier density and Hall mobility of prepared samples at RT were determined using the HCS 10 system, Linseis. To measure the output power for pure SWCNT, PPy, and SWCNT/PPy nanocomposites, rectangular (4 × 6 × 10 mm) single-leg modules were fabricated using a manual hydraulic press. These modules were fixed on a ceramic substrate, and both sides of the modules were connected to the measuring apparatus using aluminum electrodes. Strips of aluminum electrodes were used to cover the two sides. One side of the modules was placed on the hot plate that was heated to the maximum temperature (370 K). To measure the output voltage and current, a high sensitivity I–V measurement system (Keithley Instruments, Solon, OH, USA) was utilized.

3. Results and Discussion

The surface morphology and microstructure of the produced SWCNT/PPy nanocomposites were investigated by both SEM and TEM techniques, as shown in Figure 1; Figure 2. The SEM images show that the PPy was clearly formed on the sidewalls of the SWCNT, resulting in a one-dimensional (1D) nanocomposite (Figure 1). SEM images reveal that the PPy smooth coating had a considerable impact on the diameter of the SWCNT. As a systematic coating could be seen, smaller amounts of PPy could result in thinner shells/layers of PPy on the walls of SWCNT, while higher amounts of PPy could produce thicker layers of PPy. This coating was achieved by using different amounts of PPy (increasing the amount of PPy from 0.05 to 2 mL) and fixing the amount of SWCNTs, as displayed in Table 1. As shown in Figure 1a, the SWCNTs exist mostly in bundles with various diameters, and therefore, the formed core shell nanocomposites are present in various diameters (Figure 1b–f), which can be also seen clearly in the TEM images in Figure 2. From this figure, the uncoated and coated PPy-SWCNT with different PPy layer thicknesses show a similar trend to that observed in SEM images. As the amounts of PPy increase, the coating thickness increases, as shown in Figure 2a–g. Moreover, it can be clearly seen the SWCNT is situated in the cores of these 1D structures, whereas the PPy forms the shells (Figure 2h). This kind of smooth coating was also observed previously in MWCNT/PPy [25].

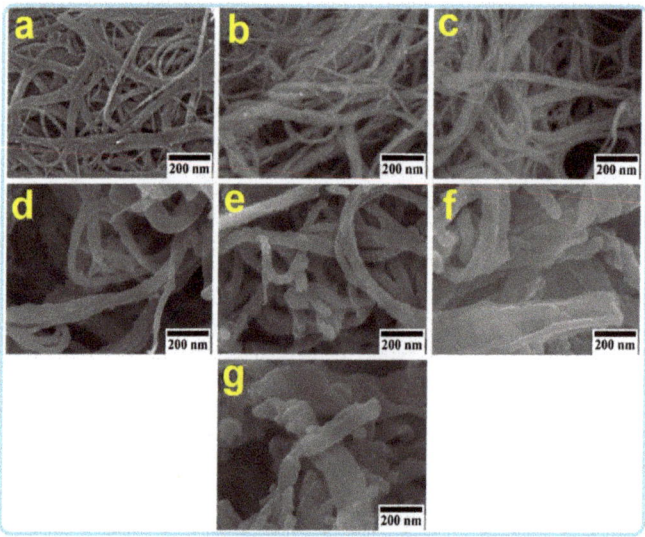

Figure 1. SEM images at the same magnification of SWCNT/PPyl nanocomposites. Images of pure SWCNT and PPy are also shown. (**a**) SWCNT, (**b**) PW1, (**c**) PW2, (**d**) PW3, (**e**) PW4, (**f**) PW5, and (**g**) PPy.

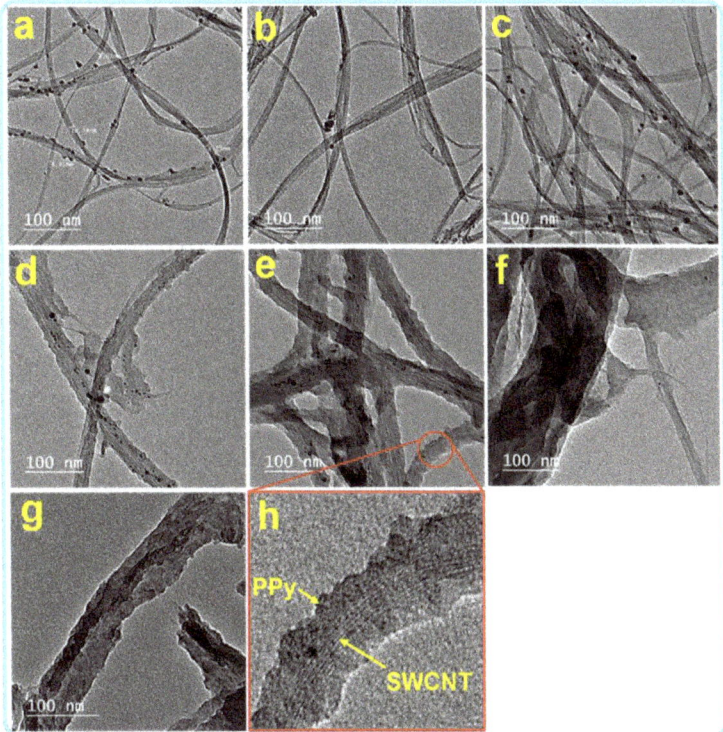

Figure 2. TEM images obtained at the same magnification for SWCNT/PPy nanocomposites. Images of pure SWCNT and PPy are also shown. (**a**) SWCNT, (**b**) PW1, (**c**) PW2, (**d**) PW3, (**e**) PW4, (**f**) PW5, (**g**) PPy. The image in (**h**) is a high-resolution TEM image for the PW4 sample.

Raman and FTIR spectra of the SWCNT, PPy, and the SWCNT/PPy core shell nanocomposites were recorded and are presented in detail in the supplementary data in Figures S1 and S2, respectively, while the results of utilizing the XRD pattern of these samples are presented in Figure S3. The elemental composition and chemical states of C1s present in the SWCNT, PPy, and SWCNT/PPy core shell nanocomposites were investigated using the XPS technique, as demonstrated in Figures S4 and S5. The XPS survey profiles of these samples are shown in Figure S4, while the band of C1s of the uncoated SWCNT, SWCNT/PPy nanocomposites, and pure PPy was deconvoluted and presented in Figure S5a–g. The results are described and interpreted in the Supplementary Materials.

The TE properties of the pure PPy, SWCNT, and SWCNT/PPy core shell nanocomposites (PW1–PW5) are presented in Figure 3a–e. At room temperature (RT), the measured electrical conductivity for pure PPy and SWCNT is approximately equal to 1330 and 113,510 S/m, respectively (Figure 3a). When the temperature increases to 350 K, the value of PPy slightly rises to 1455 S/m, while the SWCNT decreases to 104,069 S/m. PPy showed semiconductor behavior while SWCNT showed typical metallic conducting behavior. Therefore, the electrical conductivity of the produced PPy molecules is comparable to or even higher than those with similar structures or morphologies, as reported in the literature [20,35]. The electrical conductivity of the SWCNT/PPy nanocomposites with different PPy layer thicknesses is presented in Figure 3a. The obtained values were recorded as a function of temperature in the temperature range of 300–350 K. The values of coated SWCNT at RT were increased with an increase in the thickness of the coating PPy layer, mainly of the PW1and PW2 samples. The measured values at RT are equal to 123,000 and

178,000 S/m for PW1 and PW2, respectively. These values were decreased with an increase in the temperature up to 350 K, reaching 120,616 S/m for the PW1 sample and 173,637 S/m for PW2 samples, indicating degenerate semiconductor behavior. In contrast, the electrical conductivity values of the remaining samples (PW3, PW, and PW5) were decreased with the increase in thickness of the coating PPy layer. At RT, the electrical conductivity of PW3, PW4, and PW5 is approximately 68,000, 24,000, and 14,000 S/m, respectively, while at 350 K, the electrical conductivity values of these samples (PW3, PW4, and PW5) were 25–30% higher than those recorded at RT, which explains the semiconductor behavior.

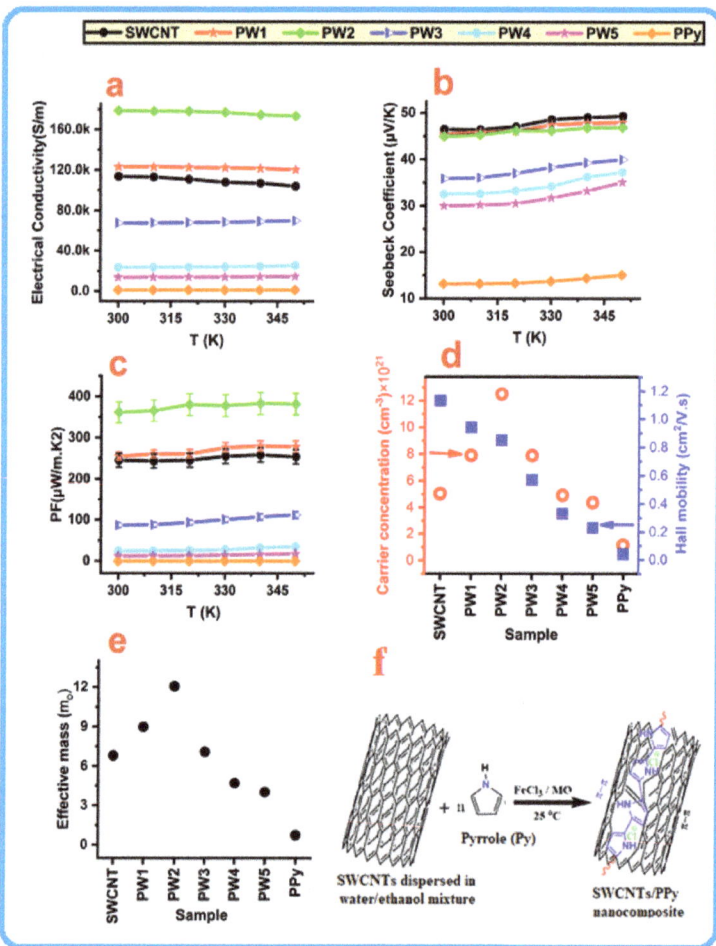

Figure 3. TE performance of the SWCNT/PPy nanocomposites as a function of the temperature (**a**–**c**). The charge carrier concentration, Hall mobility (**d**), effective mass (**e**) at RT of the pure PPy, SWCNT, and SWCNT/PPy nanocomposite samples are also shown. (**f**) is a schematic illustration showing the π–π conjugation interaction between Ppy and SWCNT during the polymerization.

This substantial improvement in the electrical conductivity of the SWCNT by Ppy coating at optimum concentration is remarkable, which is probably attributed to the smooth and flawless Ppy coating generated on the wall of SWCNT. This coating perhaps assisted charge transport and boosted the charge concentrations, simply by combining the carriers that exist in both the SWCNTs and Ppy. These findings match those of Ppy nanowire/graphene nanocomposites reported previously [35]. Furthermore, according to previous studies reported in the literature [36], the electrical conductivity of SWCNT/Ppy core/shell nanocomposites may be improved by using SWCNT as a template for the self-assembly of Ppy. This, therefore, enhances the order of crystalline alignments by the strong π–π conjugation interaction between Ppy and SWCNT during the polymerization. This interaction is shown in Figure 3f.

The measured Seebeck coefficients, as a function of temperature for the pure PPy, SWCNT, and SWCNT/PPy nanocomposites (PW1–PW5), are displayed in Figure 3b. All tested samples had a positive Seebeck coefficient, indicating a p-type conductive behavior. At 300 K, the measured Seebeck coefficients of SWCNT and PPy are equal to 46.5 µV/K and 13.2 µV/K, respectively; however, these values were slightly increased by heating the samples to 350 K. The nanocomposite samples (PW1–PW5) showed intermediate values between those of SWCNT and PPy. However, the samples of PW1 and PW2 have closer Seebeck values to that of SWCNT. These remarkable results for TE application were mainly obtained for PW2, which was also found to have the highest electrical conductivity. The thicker layers of PPy on SWCNT for PW3–PW5 significantly reduced the Seebeck values, but they were still higher than the Seebeck values of PPy.

The calculated power factor (PF) values of the pure PPy, SWCNT, and SWCNT/PPy nanocomposites were plotted as a function of temperature, and the obtained results are shown in Figure 3c. At RT, the power factor values of the uncoated SWCNT and PPy were found to be around 245 and 0.2 µW/mK^2, respectively. By excluding samples of PW1 and PW2, the calculated PF of the SWCNT/PPy nanocomposites decreases for the thicker coating layer of the PPy. The PF value of the PW2 sample is equal to 362 µW/mK^2, which is higher than values of the pure or other nanocomposite samples. This value is achieved due to the enhancement in the electrical conductivity of the PW2 sample (Figure 3a). This is likely due to the use of well-coated PPy (that have fewer defects or agglomerations) on the surface of the selected highly conducting SWCNT (Figure 2; Figure 3). The other coated samples with thicker layers of PPy (PW3–PW5) seem to contain some defects and agglomerations of PPy and also some residuals of MO molecules on the sidewalls of the SWCNT (see Figure S1). This is expected to reduce the charge transport and decrease the electrical conductivity. In general, the use of SWCNT as a template for the self-assembly of PPy will facilitate the crystalline alignments of PPy chains and make them more ordered by the strong π–π conjugation interaction between PPy and SWCNT during polymerization. This interaction will enhance the electrical conductivity by acquiring more carriers, as PPy is rich with charge carriers (see Figure 3d), but at certain layer thickness of highly pure PPy (with no defects/agglomerations or residuals). It is worth mentioning that the recorded PF value of the optimum nanocomposite sample in this study, PW2, is greater than the PF values of MWCNT/PPy and SWCNTs/PPy composites reported in the literature [36,37].

Hall effect characterization of the PPy, SWCNT, and SWCNT/PPy nanocomposite samples was performed to demonstrate the effect of PPy coating on the SWCNT in terms of electrical conductivity. Figure 3d depicts the charge carrier concentration and Hall mobility of these samples at 300 K. From the figure, the maximum carrier concentration of 12.5×10^{21} cm^{-3} was achieved for the PW2 (SWCNT/PPy (0.1)), whereas the corresponding Hall mobility was equal to 0.85 cm^2/Vs. On the other hand, the pure PPy recorded the lower charge carrier concentration of 1.1×10^{21} cm^{-3}, while the Hall mobility was around 0.05 cm^2/Vs. Moreover, the carrier concentration and the Hall mobility of SWCNT are approximately equal to 5×10^{21} cm^{-3} and 1.1 cm^2/Vs, respectively. The charge carrier concentration was increased dramatically due to the coated SWCNT with a thinner PPy layer, especially in the PW2 sample. This indicates that the carrier concentration is a crucial factor to enhance electrical conductivity by PPy coating on the SWCNT walls. The electrical conductivity value of the coated sample of PW2 was 57% higher than that value of SWCNT (Figure 3a). In general, it is noticeable that coating SWCNTs with PPy is a useful approach to enhance the electrical conductivity and similarly might be suitable for other carbon nanotubes (CNTs) [18,25].

To further elaborate on the electrical conductivity, the effective mass (m^*) values were also estimated, as displayed in Figure 3e. The above results of the electrical conductivity (Figure 3a) show that the core shell nanocomposites seem to be degenerate semiconductors, mainly for PW1 and PW2 samples; therefore, the effective mass m^* can be obtained from the Seebeck coefficient, S, and the carrier concentration, n (p for holes), according to Pisarenko's relationship [38]:

$$S = \frac{8\pi^2 k_B^2}{3eh^2} m^* T \left(\frac{\pi}{3n}\right)^{2/3} \quad (1)$$

where k_B is the Boltzmann constant = 1.38×10^{-23} m^{2}kgs^{-2}K^{-1}; h is Planck's constant = 6.63×10^{-34} m^{2}kg/s; and e = the electron charge = 1.6×10^{-19} Coulombs.

Figure 3e shows that the maximum effective mass value is equal to around 12 m_o for the PW2 (SWCNT/PPy (0.1)). The high effective mass value along with the relatively high value of mean free bath for carrier scattering might be the reasons for the electrical conductivity enhancement in this sample (PW2). This was perhaps facilitated due to the smooth coating of PPy on the SWCNT sidewalls, which enables the charge transport. On the other side, the pure PPy showed the lowest effective mass of 0.7 m_o, while the mean free path was equal to around 0.01 nm. Moreover, the effective mass of neat SWCNT was around 7 m_o. In the case of thicker PPy coatings (PW3–PW5), the effective mass values were decreased with increasing the thickness of the PPy layer. This perhaps is due to the presence of some branches or agglomerations of PPy particles within the coated layers (Figure 1; Figure 2).

The thermal conductivity (κ_{total}) of a material can be described as the sum of electronic thermal conductivity and phonon's lattice thermal conductivity. This means that thermal conductivity is the total amount of heat transferred through the material by electron/hole transporting (κ_e) and that which is transferred by phonons traveling (κ_p). The value of κ_e can be obtained using the measured electrical conductivity and then the Wiedemann–Franz law [39] on the premise that the relaxation times of phonons and electron holes are the same.

$$L = \frac{\kappa_e}{\sigma T} = \frac{\pi^2 k_B^2}{3e^2} = 2 \times 10^{-8} \text{W}\Omega\text{K}^{-2} \quad (2)$$

where L is the Lorenz number, k_B is Boltzmann's constant, and e is the electron charge. The values of κ_{total}, κ_p, and κ_e of the neat PPy, SWCNT, and SWCNT/PPy core shell nanocomposites (PW1–PW5) are presented as a function of temperature in Figure 4a–c.

At RT, the κ_{total} of SWCNT was equal to 1.08 W/mK. When the temperature was raised to 350 K, this value slightly increased to around 1.16 W/mK (Figure 4a). These values are low compared to those published in the literature [40]. This may be due to the presence of SWCNTs in bulk bundles with a complex network that has many interfaces (Figure 1a). The κ_{total} in the case of pure PPy is around 0.3 in the temperature range of room temperature to 350 K. This value is similar to those values reported in the previous study [23]. It is obvious that the κ_{total} of SWCNT appears to be almost temperature-dependent, which differs from the case of pure PPy sample. As the temperature increases, network structures may expand, resulting in close contact between nearby nanotubes of SWCNT. Consequently, it may minimize phonon scattering sites and enhance phonon transit, which therefore leads to increased thermal conductivity, κ_{total}.

The κ_p of SWCNT is small compared to its κ_{total}, while the κ_p of pure PPy is closer to its κ_{total}, as displayed in Figure 4a,b. This interestingly indicates that the electrons are the major heat carriers in the SWCNT sample, whereas the phonons are the major heat carriers in pure PPy. The κ_{total} of the SWCNT/PPy nanocomposite samples (PW1–PW5) are presented in Figure 4a. Those samples demonstrate a trend similar to that shown by electrical conductivity. Intentionally, κ_{total} increased with an increase in the wrapping layer of PPy on SWCNTs (PW1 and PW2). Then, it is decreased with an increase in the thickness of the coated layers, as can be seen in samples PW3–PW5. At RT, PW1 and PW2 samples recorded the highest thermal conductivity values of around 1.12 and 1.2 W/mK, respectively. These values slightly increased with an increase in temperature up to 350 K, reaching to 1.2 and 1.26 W/mK for samples PW1 and PW2, respectively. The κ_{total} of PW3, PW4, and PW5 samples were found to be approximately 0.9, 0.8, and 0.5 W/mK, respectively, at both RT as well as at high temperature (383 K). The κ_{total} of these samples (PW3–PW5) are lesser than those of uncoated SWCNTs. The effect of the interfacial sites between the CNTs and the PPy layer may play an important role in the scattering of more phonons. Normally, at high temperatures, the total thermal conductivity decreases due to the enhanced phonon scattering (role of κ_p), but in the present case, the observed slight increases by heating might be due to the role of κ_e, which are found to be the major heat carriers or major contributors to the total thermal conductivity in most of these samples.

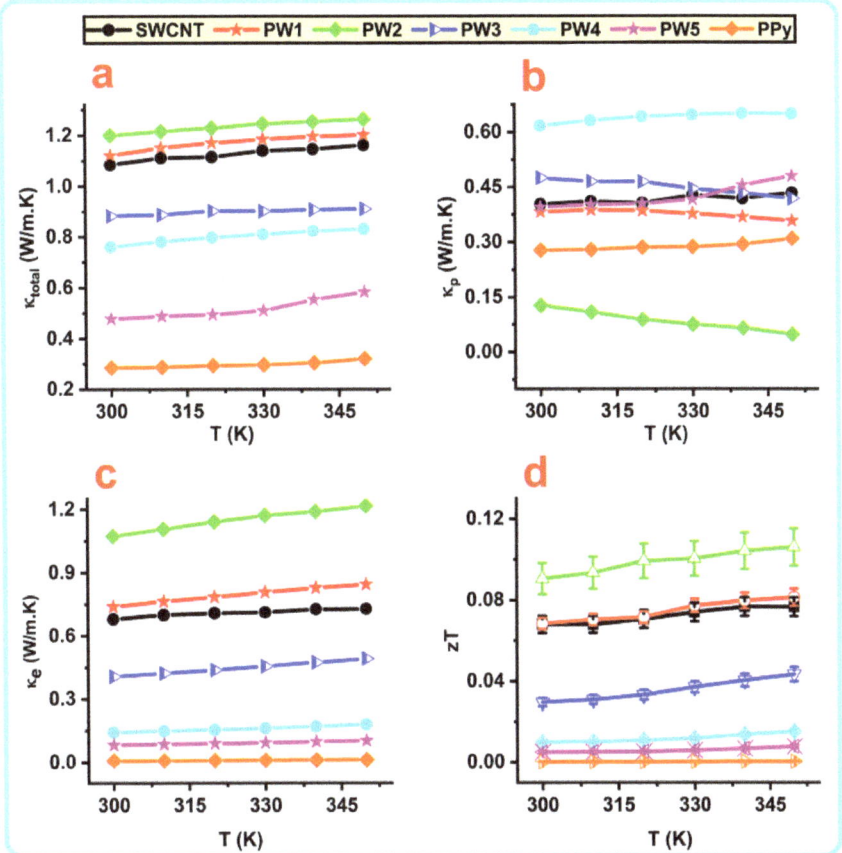

Figure 4. Total thermal conductivity, κ_{total} (**a**), phonon thermal conductivity, κ_p (**b**), electron thermal conductivity, κ_e (**c**), and figure of merit, zT (**d**) of the SWCNT/PPy core shell nanocomposites as a function of temperature.

It is clear that PW2 has the lowest value of κ_p (around 0.12 W/mK) but has the highest value of κ_e (1.07 W/mK). This indicates that electrons are the major heat carriers in this 1D nanocomposite sample. The results in Figure 3a,d show that this sample possesses the highest electrical conductivity and the maximum charge carrier concentration. The PW3 sample shows that the contributions of κ_p and κ_e are almost equal, while the κ_p of the remaining samples (PW4 and PW5) are closer to those of κ_{total}. It is clear that the κ_e contribution seems to be small in these nanocomposites, mainly in PW4 and PW5 samples. Nevertheless, the thermal conductivity of the SWCNTs coated with the optimum amount of PPy in this work is significantly lower than the values of similar composites described in the literature [20].

The assessments of figure of merit (zT) as a function of temperature for all prepared samples are displayed in Figure 4d. The zT was dramatically raised thanks to the flawless 1D coating of PPy on the sidewalls of the SWCNTs. The zT for these samples demonstrated a trend similar to that shown by electrical conductivity. At RT, the uncoated SWCNT value was equal to 0.067, while it slightly increased to 0.076 by heating to 350 K. The pure PPy exhibited significantly lower zT values than those of SWCN, with values equal to 0.5×10^{-3} at RT and 2.0×10^{-3} at 440 K. The zT values of the slightly coated SWCNTs (PW1 and PW2) were significantly higher than those of both the SWCNT and PPy. The highest zT value of 0.09 was acquired for the PW2 sample (SWCNT/PPy(0.1)) at RT and this value was increased to nearly 0.11 at 350 K.

To compare the TE performance of the present SWCNT/PPy nanocomposites in this work with those of the similar composites, Table 2 summarizes the comparison values of the electrical conductivity, Seebeck coefficient, and power factor at 300 K for the nanocomposite materials based on carbon nanotubes and PPy. From the table, it is obvious that the present SWCNT/PPy nanocomposites have superior TE properties. The optimum SWCNT/PPy nanocomposite of PW2 shows the highest PF and zT values. Moreover, this study presents a clear strategy to wrap highly conducting SWCNTs with smooth PPy layers without PPy agglomeration to form 1D core shell nanocomposite with superior TE performance. These improvements in the TE performance of the PPy-coated SWCNT nanocomposites are due to the utilization of highly conducting SWCNTs and, more particularly, this massive increase because of coating with PPy. The Seebeck coefficient is also reasonable, beside the low thermal conductivity, leading to a nanocomposite with a high zT value.

Table 2. The optimum thermoelectric properties of CNT/PPy composites.

Nanocomposite Materials	σ (S/m)	S (μV/K)	P.F. (μW/mK^2)	zT	Ref.
PPy/rGO	4160	26.9	3.01	-	[21]
PPy/GNs (PPy/graphene nanosheets)	10,168	31.74	10.24	2.80×10^{-3}	[20]
PPy/rGO thin film	8000	29	7.28	-	[41]
PPy nanowire/rGO	7500.1	33.8	8.56 ± 0.76	-	[12]
PPy/SWCNT (60 wt %)	39,900	22.2	19.7 ± 0.8	-	[19]
PPy/MWCNT (20 wt %)	~3150	~25.4	2.079	-	[36]
PPy/MWCNT (68 wt %) (at RT)	3670	24.5	2.2	-	[17]
PPy nanowire/SWCNT (60 wt %) (at RT)	~30,000	~25	21.7 ± 0.8	-	[10]
PPy/graphene (20 wt %) (at 380 K)	3690	16.6	1.01	-	[35]
SWCNT/PPy (40 wt %) (at 398 K)	10,699	22.59	5.46	-	[42]
PPy/SWCNTs film (at RT)	34,160	33.2	37.6 ± 2.3	-	[37]
MWCNTs/PPy	4000	~14	0.77	1×10^{-3}	[18]
PPy/acid-doped SWCNT	385,000	25	240.3	-	[43]
SWCNT/PPy (at RT)	**178,835**	**45**	**362**	**90×10^{-3}**	**This work**
SWCNT/PPy (at 350 K)	**173,637**	**47**	**382**	**110×10^{-3}**	

The main reasons for improving the TE properties of the SWCNT/PPy nanocomposite might be further elaborated. The first reason is using the SWCNT, which has high electrical conductivity, beside its large surface area and aspect ratio that acted as conducting bridges or conducting networks linking the PPy conducting domains [44]. The second explanation is the enhancement of the electrical conductivity of SWCNT/PPy core/shell nanocomposites, which could be increased by using SWCNT as a template for the self-assembly of PPy, which makes crystalline alignments more ordered by the strong π–π conjugation interaction between PPy and SWCNT during polymerization (Figure 3f) [45]. The third reason is the energy filtering effect at the SWCNT/PPy interfaces, where adequate potential boundary barriers positively permitted the carriers with high energy to pass, which enhances (or maintain) the Seebeck coefficient [20]. It is understood that forming core shell nanocomposites is expected to enrich the interface sites. These sites are clearly shown in the HRTEM image presented in Figure 2h. However, in the present case, such filtering effects could only maintain the Seebeck values of the PW1 and PW2 samples. This is almost in agreement with the explanation given by Neophytou et al. [46]. Moreover, since SWCNT has a complex network and the interfaces between the SWCNT/PPy nanocomposites themselves are intricate, a significant influence on phonon scattering might result in low thermal conductivity.

The presented results in this work showed a remarkable finding: SWCNTs' electrical conductivity may be considerably increased by properly wrapping them with a conducting polymer such as PPy to produce a 1D core shell nanocomposite. This might be due to the fact that conducting polymers such as PPy possess this feature by using appropriate precursors and suitable surfactants during the polarization process. It is therefore recommended to apply this approach to other highly conducting carbon materials and other conducting polymers. This straightforward procedure may serve to optimize the TE performance of TE-based polymer materials and it does not require a lengthy procedure or the use of intricate ternary composites. However, the procedure should be addressed very carefully, as sometimes this might lead to a negative result; Wang et al. [47] reported that a decrease in the electrical conductivity of other polymers might occur after adding carbon nanotubes, stating that "polymer/carbon nanotube composites exhibit poorer electrical conductivity than pure carbon nanotubes".

A thermoelectric generator (TEG) is a small and solid-state apparatus employed to convert heat energy into electricity by a phenomenon called the thermoelectric Seebeck effect. In TEG devices, several factors should be examined, including the output voltage (V), current (I), and power (P) of the single-leg module generated by the temperature gradient between the two sides of a pressed compact. In real conditions, the power generation characteristics of the single cuboid-shaped leg modules composed of SWCNT, PPy, and SWCNT/PPy nanocomposite (PW2) were examined, and the results are given in Figure 5a–f. The measured V and P as a function of I under different ΔT showed that the SWCNT/PPy nanocomposite exhibited improved TE power compared to the individual samples. The dimensions of a single cuboid leg module are presented in Figure 5g. It can be seen that the V linearly decreases as the I increases at a fixed ΔT, and this I–V linear relationship remains unaltered when ΔT is changed; therefore, Ohm's law holds true at any ΔT. Moreover, the V increased proportionally with the increase in ΔT due to the Seebeck effect. It has been also noted that the slopes of all I–V curves are nearly identical, demonstrating that ΔT has no major effect on the internal resistance (R) of all samples. As ΔT increases, the output power also increases; at $\Delta T = 40$ K, the P_{max} of 24.42 nW (the equivalent power density is ~61 nW/cm^2) was achieved for the PW2 sample. The experimental results were also validated by generating numerical simulation curves (Figure 5a–f). These curves were generated by a three-dimensional ANSYS numerical model with 20 node coupling elements, SOLID226, selected to simulate the thermoelectric current and voltage, similar to the work reported in the literature [48–51]. More details about this model are described in the Supplementary Data. From the figure, it can be concluded that the numerical simulation curves are in agreement with the experimental results.

The above findings show that the TEG of SWCNT/PPy(0.1) nanocomposites is superior compared to the individual samples, e.g., SWCNT and PPy. The attained findings are consistent with the PF and zT values of each leg, as shown in Figures 3c and 4d. The PPy coating significantly improved the TE power output of the SWCNT and served as an excellent binder for the SWCNT, which is required for fabricating a strong leg. The results presented above concern only one leg of the SWCNT wrapped with a PPy semiconductor. Utilizing a large number of p–n pairs with acceptable matching might be used to create a functional electrical device with greater TE power generation.

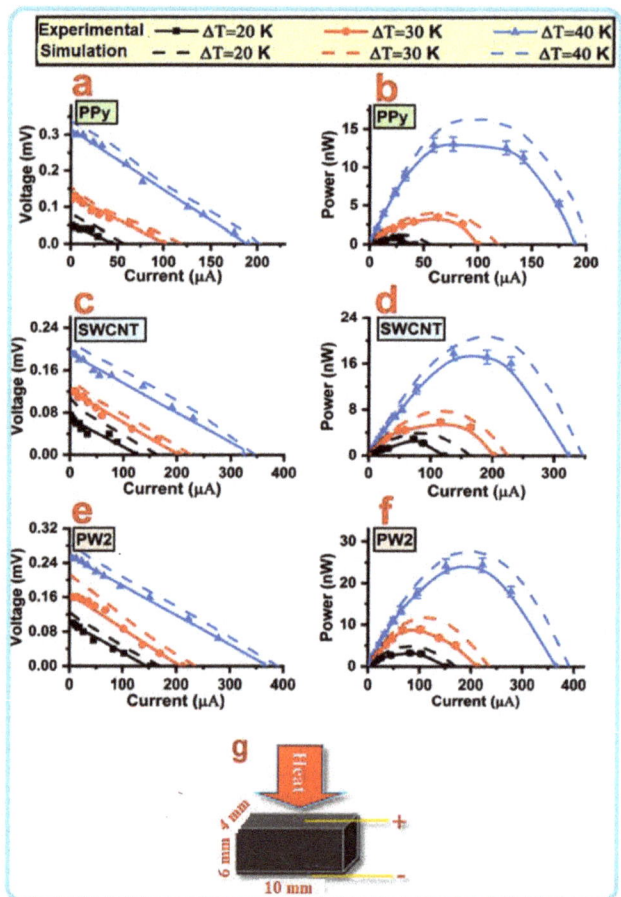

Figure 5. TE power generation characteristics of the single-leg module of the SWCNT/PPy core shell nanocomposite of sample PW2. The same for pure SWCNT and PPy are also shown for comparison. The simulated curves performed by ANSYS software are also shown (**a–f**). A single squared leg module is shown in (**g**).

4. Conclusions

In this work, a highly conducting SWCNT was combined with PPy as a 1D core shell nanocomposite for higher TE power generation. Various amounts of Py as a monomer source for PPy were coated on SWCNT along with methyl orange as a surfactant and ferric chloride as an initiator. The TE power factor and figure of merit values of the optimized core shell nanocomposite sample of SWCNT/PPy were greatly increased to 360 µW/mK^2 and 0.09, respectively, which demonstrate significantly better TE performance than the individual materials. The power generation characteristics of a single-leg module of the optimized sample confirmed these enhancements. This enhancement was attributed to the uniform coating and good interaction between PPy polymer chains and walls of the SWCNT through π–π stacking. The achieved enhancement in the TE performance of SWCNT/PPy nanocomposite is found to be superior to those reported of similar composites based on carbon and PPy materials. It is therefore of importance to continue developing such nanocomposites that might be recommended as a scalable TE material for the future of generating high TE power.

Supplementary Materials: The following supporting information can be downloaded at: https://www.mdpi.com/article/10.3390/nano12152582/s1.

Author Contributions: M.A. (M. Almasoudi) and N.S. wrote the paper; M.A. and A.S. carried out the thermal conductivity measurements and analysis; A.A. measured the TE performance of the samples; M.S.Z. and M.H.A.-A. designed the polymer experiments; N.S. and K.K. analyzed the data and revised the manuscript; M.A. (Mutabe Aljaghtham) generated the numerical simulation curves. All authors have read and agreed to the published version of the manuscript.

Funding: This research was funded by King Abdullah City for Atomic and Renewable Energy, KSA.

Data Availability Statement: The data supporting the findings of this study are available upon request to nsalah@kau.edu.sa.

Acknowledgments: The authors acknowledge the support provided by King Abdullah City for Atomic and Renewable Energy (K.A.CARE) under the K.A.CARE-King Abdulaziz University Collaboration Program.

Conflicts of Interest: The authors declare no conflict of interest.

References

1. Gao, C.; Chen, G. Conducting polymer/carbon particle thermoelectric composites: Emerging green energy materials. *Compos. Sci. Technol.* **2016**, *124*, 52–70. [CrossRef]
2. Glaudell, A.M.; Cochran, J.E.; Patel, S.N. Impact of the doping method on conductivity and thermopower in semiconducting polythiophenes. *Adv. Energy Mater.* **2015**, *5*, 1401072. [CrossRef]
3. Krause, B.; Liguoro, A.; Pötschke, P. Blend Structure and n-Type Thermoelectric Performance of PA6/SAN and PA6/PMMA Blends Filled with Singlewalled Carbon Nanotubes. *Nanomaterials* **2021**, *11*, 1146. [CrossRef] [PubMed]
4. Liu, S.; Li, H.; Li, P.; Liu, Y.; He, C. Recent Advances in Polyaniline-Based Thermoelectric Composites. *CCS Chem.* **2021**, *3*, 2547–2560. [CrossRef]
5. Li, C.; Ma, H.; Tian, Z. Thermoelectric properties of crystalline and amorphous polypyrrole: A computational study. *Appl. Therm. Eng.* **2017**, *111*, 1441–1447. [CrossRef]
6. Tonga, M.; Wei, L.; Lahti, P.M. Enhanced thermoelectric properties of PEDOT:PSS composites by functionalized single wall carbon nanotubes. *Int. J. Energ. Res.* **2020**, *44*, 9149–9156. [CrossRef]
7. Jiang, Q.; Liu, C.; Xu, J.; Lu, B.; Song, H.; Shi, H.; Yao, Y.; Zhang, L. An effective substrate for the enhancement of thermoelectric properties in PEDOT: PSS. *J. Polym. Sci. B Polym. Phys.* **2014**, *52*, 737–742. [CrossRef]
8. Xu, K.; Gao, C.; Chen, G.; Qiu, D. Direct evidence for effect of molecular orientation on thermoelectric performance of organic polymer materials by infrared dichroism. *Org. Electron.* **2016**, *31*, 41–47. [CrossRef]
9. Liang, L.; Chen, G.; Guo, C.-Y. Polypyrrole nanostructures and their thermoelectric performance. *Mater. Chem. Front.* **2017**, *1*, 380–386. [CrossRef]
10. Liang, L.; Chen, G.; Guo, C.-Y. Enhanced thermoelectric performance by self-assembled layered morphology of polypyrrole nanowire/single-walled carbon nanotube composites. *Compos. Sci. Technol.* **2016**, *129*, 130–136. [CrossRef]
11. Zhang, Z.; Chen, G.; Wang, H.; Li, X. Template-directed in situ polymerization preparation of nanocomposites of PEDOT:PSS-coated multi-walled carbon nanotubes with enhanced thermoelectric property. *Chem. Asian J.* **2015**, *10*, 149–153. [CrossRef]
12. Zhang, Z.; Chen, G.; Wang, H.; Zhai, W. Enhanced thermoelectric property by the construction of a nanocomposite 3D interconnected architecture consisting of graphene nanolayers sandwiched by polypyrrole nanowires. *J. Mater. Chem. C* **2015**, *3*, 1649–1654. [CrossRef]
13. Li, J.; Du, Y.; Jia, R.; Xu, J.; Shen, S.Z. Flexible thermoelectric composite films of polypyrrole nanotubes coated paper. *Coatings* **2017**, *7*, 211. [CrossRef]
14. Wu, J.; Sun, Y.; Pei, W.-B.; Huang, L.; Xu, W.; Zhang, Q. Polypyrrole nanotube film for flexible thermoelectric application. *Synth. Met.* **2014**, *196*, 173–177. [CrossRef]
15. Culebras, M.; Uriol, B.; Gómez, C.M.; Cantarero, A. Controlling the thermoelectric properties of polymers: Application to PEDOT and polypyrrole. *Phys. Chem. Chem. Phys.* **2015**, *17*, 15140–15145. [CrossRef]
16. Zhang, W.-D.; Xiao, H.-M.; Fu, S.-Y. Preparation and characterization of novel polypyrrole-nanotube/polyaniline free-standing composite films via facile solvent-evaporation method. *Compos. Sci. Technol.* **2012**, *72*, 1812–1817. [CrossRef]
17. Song, H.; Cai, K.; Wang, J.; Shen, S. Influence of polymerization method on the thermoelectric properties of multi-walled carbon nanotubes/polypyrrole composites. *Synth. Met.* **2016**, *211*, 58–65. [CrossRef]
18. Baghdadi, N.; Zoromba, M.S.; Abdel-Aziz, M.H.; Al-Hossainy, A.F.; Bassyouni, M.; Salah, N. One-Dimensional Nanocomposites Based on Polypyrrole-Carbon Nanotubes and Their Thermoelectric Performance. *Polymers* **2021**, *13*, 278. [CrossRef]
19. Liang, L.; Gao, C.; Chen, G.; Guo, C.-Y. Large-area, stretchable, super flexible and mechanically stable thermoelectric films of polymer/carbon nanotube composites. *J. Mater. Chem. C* **2016**, *4*, 526–532. [CrossRef]

20. Wang, L.; Liu, F.; Jin, C.; Zhang, T.; Yin, Q. Preparation of polypyrrole/graphene nanosheets composites with enhanced thermoelectric properties. *RSC Adv.* **2014**, *4*, 46187–46193. [CrossRef]
21. Han, S.; Zhai, W.; Chen, G.; Wang, X. Morphology and thermoelectric properties of graphene nanosheets enwrapped with polypyrrole. *RSC Adv.* **2014**, *4*, 29281–29285. [CrossRef]
22. Zoromba, M.S.; Abdel-Aziz, M.H.; Bassyouni, M.; Abusorrah, A.M.; Attar, A.; Baghdadi, N.; Salah, N. Polypyrrole sheets composed of nanoparticles as a promising room temperature thermo-electric material. *Phys. E* **2021**, *134*, 114889. [CrossRef]
23. Almasoudi, M.; Zoromba, M.S.; Abdel-Aziz, M.H.; Bassyouni, M.; Alshahrie, A.; Abusorrah, A.M.; Salah, N. Optimization preparation of one-dimensional polypyrrole nanotubes for enhanced thermoelectric performance. *Polymer* **2021**, *225*, 123950. [CrossRef]
24. Mei, H.; Cheng, Y. Research progress of electrical properties based on carbon nanotubes; interconnection. *Ferroelectrics* **2020**, *564*, 1–18. [CrossRef]
25. Salah, N.; Alhebshi, N.A.; Salah, Y.N.; Alshareef, H.N.; Koumoto, K. Thermoelectric properties of oil fly ash-derived carbon nanotubes coated with polypyrrole. *J. Appl. Phys.* **2020**, *128*, 235104. [CrossRef]
26. Fan, X.; Yang, Z.; He, N. Hierarchical nanostructured polypyrrole/graphene composites as supercapacitor electrode. *RSC Adv.* **2015**, *5*, 15096–15102. [CrossRef]
27. Šetka, M.; Calavia, R.; Vojkůvka, L.; Liobet, E.; Drbohlavova, J.; Vallejos, S. Raman and XPS studies of ammonia sensitive polypyrrole nanorods and nanoparticles. *Sci. Rep.* **2019**, *9*, 8465. [CrossRef]
28. Zhang, B.; Xu, Y.; Zheng, Y.; Dai, L.; Zhang, M.; Yang, J.; Chen, Y.; Chen, X.; Zhou, J. A Facile Synthesis of Polypyrrole/Carbon Nanotube Composites with Ultrathin, Uniform and Thickness-Tunable Polypyrrole Shells. *Nanoscale Res. Lett.* **2011**, *6*, 431. [CrossRef]
29. Wu, M.-C.; Lin, M.-P.; Chen, S.-W.; Lee, P.-H.; Li, J.-H.; Su, W.-F. Surface-enhanced Raman scattering substrate based on a Ag coated monolayer array of SiO$_2$ spheres for organic dye detection. *RSC Adv.* **2014**, *4*, 10043–10050. [CrossRef]
30. Blinova, N.V.; Stejskal, J.; Trchová, M.; Prokeš, J.; Omastová, M. Polyaniline and polypyrrole: A comparative study of the preparation. *Eur. Polym. J.* **2007**, *43*, 2331–2341. [CrossRef]
31. Chougule, M.A.; Pawar, S.G.; Godse, P.R.; Mulik, R.N.; Sen, S.; Patil, V.B. Synthesis and characterization of polypyrrole (PPy) thin films. *Soft Nanosci. Lett.* **2011**, *1*, 6–10. [CrossRef]
32. Rafique, S.; Sharif, R.; Rashid, I.; Chani, S. Facile fabrication of novel silver-polypyrrole-multiwall carbon nanotubes nanocomposite for replacement of platinum in dye-sensitized solar cell. *AIP Adv.* **2016**, *6*, 085018. [CrossRef]
33. Márquez-Herrera, A.; Ovando-Medina, V.M.; Castillo-Reyes, B.E.; Zapata-Torres, M.; Meléndez-Lira, M.; González-Castañeda, J. Facile synthesis of SrCO$_3$-Sr(OH)$_2$/PPy nanocomposite with enhanced photocatalytic activity under visible light. *Materials* **2016**, *9*, 30. [CrossRef]
34. Bagai, R.; Christopher, J.; Kapur, G.S. Evaluating industrial grade functionalized multiwalled carbon nanotubes by X-ray photoelectron spectroscopy. *Fuller. Nanotub. Carbon Nanostruct.* **2019**, *27*, 240–246. [CrossRef]
35. Du, Y.; Niu, H.; Li, J.; Dou, Y.; Shen, S.Z.; Jia, R.; Xu, J. Morphologies tuning of polypyrrole and thermoelectric properties of polypyrrole nanowire/graphene composites. *Polymers* **2018**, *10*, 1143. [CrossRef]
36. Wang, J.; Cai, K.; Shen, S.; Dou, Y.; Shen, S.Z.; Jia, R.; Xu, J. Preparation and thermoelectric properties of multi-walled carbon nanotubes/polypyrrole composites. *Synth. Met.* **2014**, *195*, 132–136. [CrossRef]
37. Liu, Z.; Sun, J.; Song, H.; Pan, Y.; Song, Y.; Zhu, Y.; Yao, Y.; Huang, F.; Zuo, C. High performance polypyrrole/SWCNTs composite film as a promising organic thermoelectric material. *RSC Adv.* **2021**, *11*, 17704–17709. [CrossRef]
38. Govindaraj, P.; Sivasamy, M.; Murugan, K.; Venugopal, K.; Veluswamy, P. Pressure-driven thermoelectric properties of defect chalcopyrite structured ZnGa$_2$Te$_4$: Ab initio study. *RSC Adv.* **2022**, *12*, 12573–12582. [CrossRef]
39. Zhao, L.-D.; Lo, S.-H.; He, J.; Li, H.; Biswas, K.; Androulakis, J.; Wu, C.-I.; Hogan, T.P.; Chung, D.-Y.; Dravid, V.P.; et al. High performance thermoelectrics from earth-abundant materials: Enhanced figure of merit in PbS by second phase nanostructures. *J. Am. Chem. Soc.* **2011**, *133*, 20476–20487. [CrossRef]
40. Salah, N.; Baghdadi, N.; Alshahrie, A.; Saeed, A.; Ansari, A.R.; Memic, A.; Koumoto, K. Nanocomposites of CuO/SWCNT: Promising thermoelectric materials for mid-temperature thermoelectric generators. *J. Eur. Ceram. Soc.* **2019**, *39*, 3307–3314. [CrossRef]
41. Xin, S.; Yang, N.; Gao, F.; Zhao, J.; Li, L.; Teng, C. Free-standing and flexible polypyrrole nanotube/reduced graphene oxide hybrid film with promising thermoelectric performance. *Mater. Chem. Phys.* **2018**, *212*, 440–445. [CrossRef]
42. Du, Y.; Xu, J.; Lin, T. Single-walled carbon nanotube/polypyrrole thermoelectric composite materials. In Proceedings of the IOP Conference Series: Earth and Environmental Science, Banda Aceh, Indonesia, 26–27 September 2018; IOP Publishing: Bristol, UK, 2018.
43. Fan, W.; Zhang, Y.; Guo, C.-Y.; Chen, G. Toward high thermoelectric performance for polypyrrole composites by dynamic 3-phase interfacial electropolymerization and chemical doping of carbon nanotubes. *Compos. Sci. Technol.* **2019**, *183*, 107794. [CrossRef]
44. Rahaman, M.; Aldalbahi, A.; Almoiqli, M.; Alzahly, S. Chemical and electrochemical synthesis of polypyrrole using carrageenan as a dopant: Polypyrrole/multi-walled carbon nanotube nanocomposites. *Polymers* **2018**, *10*, 632. [CrossRef]
45. Luo, J.; Jiang, S.; We, Y.; Chen, M.; Liu, X. Synthesis of stable aqueous dispersion of graphene/polyaniline composite mediated by polystyrene sulfonic acid. *J. Polym. Sci. A Polym. Chem.* **2012**, *50*, 4888–4894. [CrossRef]

46. Neophytou, N.; Vargiamidis, V.; Foster, S.; Graziosi, P.; Oliveira, L.d.S.; Chakraborty, D.; Li, Z.; Thesberg, M.; Kosina, H.; Bennett, N.; et al. Hierarchically nanostructured thermoelectric materials: Challenges and opportunities for improved power factors. *Eur. Phys. J. B* **2020**, *93*, 213. [CrossRef]
47. Wang, S.; Zhou, Y.; Liu, Y.; Wang, L.; Gao, C. Enhanced thermoelectric properties of polyaniline/polypyrrole/carbon nanotube ternary composites by treatment with a secondary dopant using ferric chloride. *J. Mater. Chem. C* **2020**, *8*, 528–535. [CrossRef]
48. Angrist, S.W. *Direct Energy Conversion*, 3rd ed.; Allyn and Bacon: Boston, MA, USA, 1976.
49. Madenci, E.; Guven, I. *The Finite Element Method and Applications in Engineering Using ANSYS®*; Springer: Berlin/Heidelberg, Germany, 2015.
50. Aljaghtham, M.; Celik, E. Design optimization of oil pan thermoelectric generator to recover waste heat from internal combustion engines. *Energy* **2020**, *200*, 117547. [CrossRef]
51. Aljaghtham, M.; Celik, E. Numerical analysis of energy conversion efficiency and thermal reliability of novel, unileg segmented thermoelectric generation systems. *Int. J. Energy Res.* **2021**, *45*, 8810–8823. [CrossRef]

Article

Study on the Characteristics of the Dispersion and Conductivity of Surfactants for the Nanofluids

Sedong Kim

German Engineering Research and Development Center LSTME Busan Branch, Busan 46742, Korea; sedong.kim@lstme.org

Abstract: Given the importance of nanofluid dispersion and stability, a number of approaches were proposed and applied to the nanofluid preparation process. Among these approaches, the noncovalent chemical process was intensively utilized because of its effective dispersion ability. For the noncovalent dispersion method, polymers and surfactants are typically used. In order to find an effective noncovalent dispersion method, several types of solutions were prepared in this study. The widely used naturally cellulose nanocrystal (CNC) aqueous solution was compared with several surfactant aqueous solutions. The dispersion characteristics of the prepared fluids were examined by UV/VIS spectroscopy at operating wavelengths ranging from 190 to 500 nm. Furthermore, the heat capacity and the electrical and thermal conductivity of the fluids were analyzed to evaluate their heat transfer performance and conductivity. The Lambda system was utilized for thermal conductivity measurement with operation at proper temperature ranges. The electrical conductivity of the fluids was measured by a conductivity meter. This experimental study revealed that the cellulose nanocrystal was an effective source of the noncovalent dispersion agent for thermal characteristics and was more eco-friendly than other surfactants. Moreover, cellulose aqueous solution can be used as a highly thermal efficient base fluid for nanofluid preparation.

Keywords: nanofluid; surfactant; heat transfer performance; dispersion; cellulose

Citation: Kim, S. Study on the Characteristics of the Dispersion and Conductivity of Surfactants for the Nanofluids. *Nanomaterials* 2022, *12*, 1537. https://doi.org/10.3390/nano12091537

Academic Editors: Ting Zhang and Peng Jiang

Received: 14 April 2022
Accepted: 29 April 2022
Published: 2 May 2022

Publisher's Note: MDPI stays neutral with regard to jurisdictional claims in published maps and institutional affiliations.

Copyright: © 2022 by the author. Licensee MDPI, Basel, Switzerland. This article is an open access article distributed under the terms and conditions of the Creative Commons Attribution (CC BY) license (https://creativecommons.org/licenses/by/4.0/).

1. Introduction

Nanofluid has been studied over the past decades and has a great potential to improve heat transfer properties [1]. However, unless the nanoparticles of the nanofluid are well dispersed, enhanced thermal performance [2,3] cannot be expected [4]. For this reason, two methods are generally applied for the well dispersion of nanofluids [5]; one is a mechanical method, and the other is a chemical method. However, the ultrasonic excitation and grinding of the mechanical method can leave too many fragments in a nanofluid when CNTs are used as nanoparticles. Therefore, it is not good to reduce the aspect ratio for the degree of dispersion stability [6]. This mechanical method is time-consuming and inefficient.

Unlike the mechanical method, covalent and non-covalent chemical methods [7–10] can avoid the aggregation of nanoparticles. The covalent methods have been functionalization with many chemical moieties to enhance the solubility of solvents; however, strong chemical synthesis at high temperatures causes defects on the carbon nanotube (CNT) surface, changing the electrical characteristics of CNT. On the other hand, the noncovalent method involves the adsorption of the chemical moieties onto the CNT surface, either through π-π stacking interactions, such as in DNA; uncharged surfactants; or the Coulomb attraction in the case of charged chemical moieties. The noncovalent method is effective in the sense that it does not alter the π-electron cloud of graphene, in turn protecting the electrical characteristics of nanotubes. For instance, polymers and surfactants are widely used for CNT dispersion through non-covalent methods.

Among those for CNT dispersion, cellulose is the richest polymer in nature. Over the past few years, the development of cellulose-based materials for CNT dispersion has been reported [11,12]. Cellulose nanocrystals (CNCs) have become widespread in many studies [13,14]. They are rod-shaped nanoparticles obtained from the acid hydrolysis of cellulose. They are about 10 to 100 nm in diameter and 100 to 1000 nm in length, depending on the cellulose source and the hydrolysis conditions.

CNC strongly interacts with the water molecules through hydrogen bonding because of the hydroxyl groups on the cellulose molecules [15,16]. Surfactants are classified into cationic, anionic, nonionic, and amphoteric depending on the charge of the head group [17–20]. In ionic surfactants, the surfaces of the particles have the same charge, which can generate electrostatic repulsion. Strong electrostatic repulsion between nanoparticles promotes the stable dispersion of nanofluid [21].

Many kinds of surfactants are used with nanoparticle suspensions, such as sodium dodecyl sulfate (SDS), sodium dodecyl benzenesulfonate (SDBS), and lauryl betaine (LB), to increase the dispersion and stability of nanofluids [22,23]. In order to choose the proper surfactant for a particular application, it is very important to conduct a systematic study of different parameters, such as stability and concentration [24–28]. Yurekli et al. [29] and Hertel et al. [30] studied changes in the phase behavior of CNTs on the basis of the concentration and the type of interaction of the surfactants. However, there have been fewer systematic studies on the different, proper parameters of cellulose [31–33] influencing the nanoparticle dispersion of a base fluid.

In this study, the noncovalent process was conducted to maintain conductivity and enhance the nanoparticle dispersion for a base fluid, and [34,35] the heat capacity was also compared with that of other surfactants according to previous studies. It was found that cellulose is not only hydrophilic and eco-friendly, but it is also more thermally efficient than other surfactants. From these results, it can be confirmed that this study will make a significant contribution to the heat transfer technologies related to nanofluids because it shows good dispersion and stability to avoid nanoparticle agglomeration and identifies the stable thermal and electrical characteristics of a base fluid.

2. Materials and Methods

2.1. Materials

Demineralized water (DW) was prepared using a membrane-type DW device, which can produce DW with water quality of under 10 ppm of total dissolved solids (TDS). The cellulose nanocrystal (CNC) used in this research was extracted from the western hemlock plant and was supplied by SKB Tech, South Korea. Sodium dodecyl sulfate (SDS, $CH_3(CH_2)11OSO_3Na$) with a 288.38 relative molecular mass (Junsei Chemical Co., Ltd, Tokyo, Japan); sodium dodecyl benzene sulfonate (SDBS, $C_{18}H_{29}NaO_3S$), hard type, with a 348.48 relative molecular mass (Chemical Industry Co., Ltd, Tokyo, Japan), and dodecyl betaine (DB, $C_{16}H_{33}NO_2$) with a 271.44 relative molecular mass (Avention Co., Ltd., Seoul, Korea) were used for the dispersants.

2.2. Methods

The electrical conductivity meter (Model CM-25R) used a contacting-type conductivity sensor, which consists of an electrode. The titanium–palladium alloy electrode was specifically sized and spaced to provide a known "cell constant". A UV/Visible Spectrophotometer (X-ma 3000 Series Spectrophotometer, Human Co., Ltd., Seoul, Korea) was used to measure the dispersibility of the aqueous solution. The LAMBDA system measured the thermal conductivity of the aqueous fluids according to ASTM D 2717 by hot wire methods (evaluation of thermal conductivity of liquids). The heat capacities of all samples were measured by applying a constant heat source at the same time, and a comparison of the temperature of each surfactant was made. For all the measurements, tests were recorded at the same time, and each was carried out three times in the same way. In the repeated measurements, no significant differences were found.

2.3. Preparation of Samples

Garg et al. [36] experimentally studied the sedimentation and dispersion properties of nanofluid dispersed in a multi-walled carbon nanotube (MWCNT) over time. They reported that nanofluid treated with ultrasonic treatment for 40 min had a maximum thermal conductivity improvement effect.

Figure 1 shows the sedimentation of the MWCNT nanofluid [37]. Following Garg et al., the cellulose and surfactants were dispersed with DW, and the excitation frequency and period were 42 kHz and 40 min, respectively. CNC was heated to temperatures ranging from 50 to 90 °C because CNC solution is not easily dispersed well without heating according to Molnes et al. [38]. All samples were prepared at a concentration of 0.1 wt%.

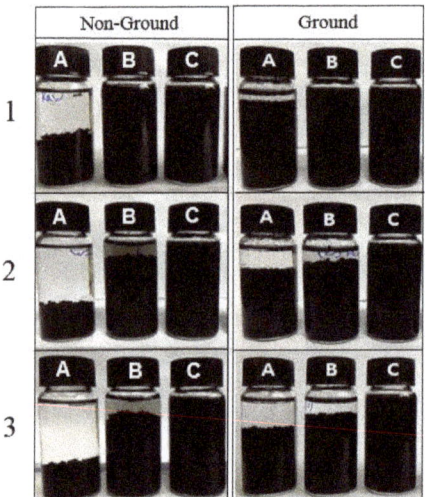

Figure 1. Sedimentation of the CNT nanofluid over time. (1) After sonication, (2) after 7 days, (3) after 30 days. Reprinted from ref. [37].

3. Results and Discussion

3.1. Structures of CNC

As seen in Figure 2, the morphological analysis of the cellulose was performed by Transmission Electron Microscopy (TEM, Technai 128 FEI). Figure 2A shows the agglomerations of nanocellulose formed as bundles by the strong hydrogen bonds between the single cellulose crystallites. However, Figure 2B,C show that the agglomerated particles consist of multiple single particles, which gathered together and formed the large aggregates. Through the TEM investigation, it can be seen that nanocellulose was formed by the aggregation of small rods.

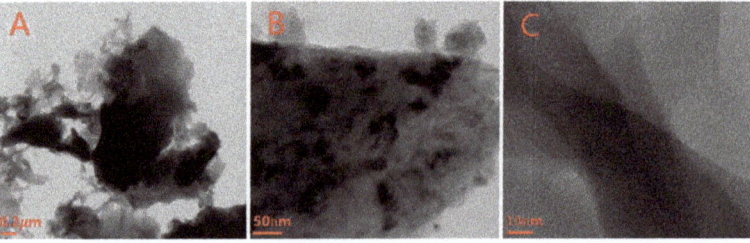

Figure 2. TEM image of cellulose nanocrystal ((**A**): 0.2 μm, (**B**): 50 nm, (**C**): 10 nm).

3.2. Electrical Conductivity of Solution

In the area of electromechanical microelectronics, the property of electrical conductivity is an important factor. Some devices need electrical conductivity, while other devices do not need conductivity because of electrical interference. For instance, it is required for the probe to have a high thermal conductivity but low electrical conductivity.

Therefore, this study investigated the electrical conductivity of several factors. The calibration was performed with a potassium chloride standard (1.41, 12.86 μS/cm) solution before the measurement.

As seen in Figure 3, the electrical conductivity of the CNC solution was lower than that of other surfactants. In the case of CNC, its electrical conductivity was not severely changed with the temperature variation, as shown in Figure 4, which means that CNC has the characteristic of low electrical conductivity regardless of the temperature variation.

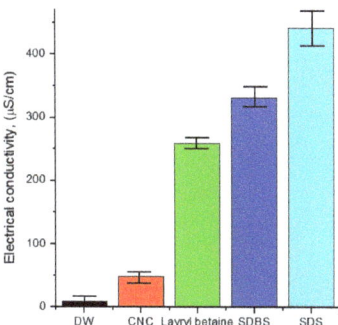

Figure 3. Electrical conductivity of each surfactant.

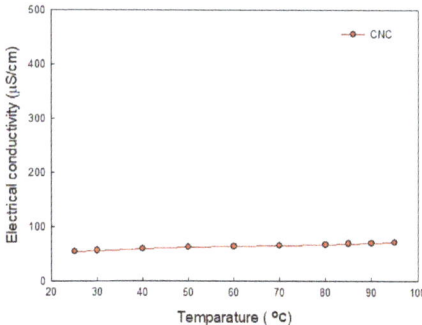

Figure 4. Electrical conductivity of CNC with increase in temperature.

In general, as the temperature increases, the electrical conductivity also increases in an aqueous solution because of fast ion diffusion. Therefore, most insulators can be combined with CNC regardless of temperature. It can be also applied in industrial fields in which the electrical conductivity of a working fluid can potentially be a threat to a system and its surroundings.

3.3. Dispersibility of Solution

In order to investigate the light absorbance characteristic of a solution in accordance with wavelength, a UV/VIS spectrophotometer was used. The UV/VIS spectrophotometer can measure the light absorbance of wavelengths ranging from 190 to 500 nm, which shows the light absorbance characteristics of the solutions.

The UV/VIS spectrophotometer is generally used for two purposes in nanofluid studies, which are the investigations of the concentration and the dispersion of nanoparticles. As the concentration of nanoparticles increases, the light absorbance increases. Additionally, the better the dispersion, the higher the light absorbance if the concentration of nanoparticles is the same.

Figure 5 presents the experimental results showing the light absorbance characteristics through the UV/VIS when each material was independently mixed with DW. Since each of the three surfactants and CNC were mixed with DW alone, Figure 4 provides the information on how the three surfactants affected the light absorbance when they were used for a well dispersion of nanoparticles.

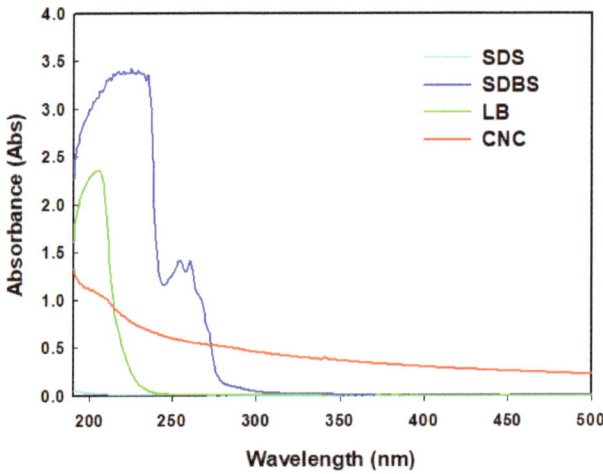

Figure 5. UV/VIS spectra of each surfactant.

Figure 5 shows that SDS does not absorb the lights over the measurement range of UV/VIS, which means that SDS is transparent in a solution and that SDS does not affect the light absorbance of a nanofluid. In the cases of SDBS and LB, some peaks were observed from 200 to 280 nm; however, those peaks fell into the range of ultraviolet rays. Therefore, they are also transparent when mixed with DW. Therefore, if three surfactants are used for a better dispersion of nanoparticles, the investigation of the visible light absorbance can be the index of a well dispersion, which means that the well dispersion can be roughly evaluated with the naked eye.

When CNC was independently mixed with DW, its light absorbance gradually decreased as the wavelength increased. Especially, it should be noticed that CNC has very low absorbance characteristics in the ultraviolet range compared with other surfactants, such as SDBS and LB. These results have an important physical meaning when making nanofluids.

The CNC has smooth absorbance characteristics over the whole UV/VIS wavelength, which indicated that CNC could provide more detailed information on the dispersion and concentration. If SDBS and LB are used as the surfactants for the dispersion, it is difficult to use light absorbance in the ultraviolet range for the evaluation criteria of dispersion and concentration. Figure 6 shows the UV/VIS measurement for the Al_2O_3 nanofluid, which was previously performed by the author [5]. As seen in Figure 6, the smooth absorbance clearly provides the concentration variation, which means that CNC is a good surfactant to evaluate the dispersion and concentration of nanoparticles.

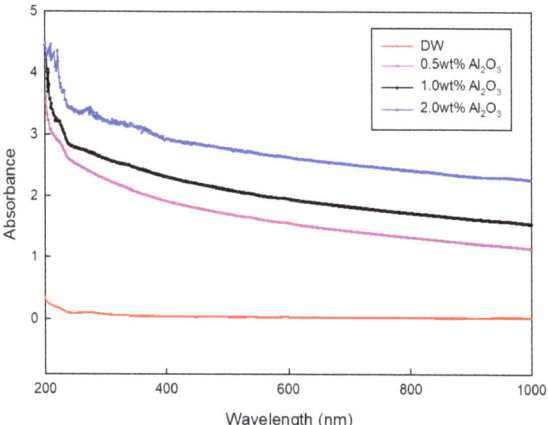

Figure 6. Absorbance of a base liquid (DW) and the nanofluids with three different alumina concentrations. Reprinted from ref. [5].

3.4. Thermal Conductivity of Solution

The thermal conductivities were measured by the LAMBDA measuring system to investigate each solution.

The LAMBDA system is based on the hot-wire method, and a platinum wire with a 0.1 mm diameter was applied as the hot wire. The detailed principles of this can be found in the previous research [39].

Figure 7 shows the thermal conductivities according to the concentration of CNC and the temperature variation. It can be seen that the thermal conductivity of CNC nanofluid decreased when the content of cellulose increased. It was reasonable to compare each surfactant at a concentration of 0.1% because when the CNC concentration was 0.1%, the value of thermal conductivity was high. Therefore, all samples were set at the proper concentration of nanofluid of 0.1 wt%.

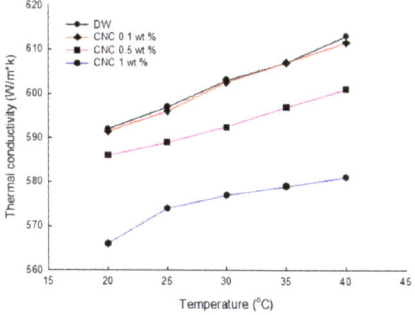

Figure 7. Thermal conductivity measurements of CNC according to concentration.

As seen in Figure 8, CNC showed the highest thermal conductivity, followed by LB, and SDS and SDBS had similar values of thermal conductivity. Generally, when a surfactant is used in the preparation of a nanofluid, the thermal conductivity is degraded [15], and thus it may be seen that a good thermal conductivity value of cellulose is excellent for the preparation of a nanofluid.

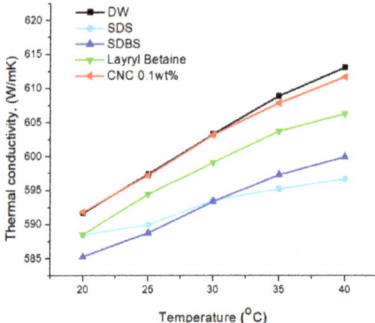

Figure 8. Thermal conductivity measurements of each surfactant.

3.5. Heat Capacity of Solution

Each sample was prepared for the same amount of solution and was constantly heated by a hot plate and magnetic stirrer, and the liquid temperatures were measured by thermally insulated T-type thermocouples. Each sample was measured for 20 min at the same initial temperature and same magnetic RPM. Temperatures were recorded in the data logger.

Figure 9 shows the result of the temperature variation of samples during the heating process. It is known that the heat capacity C of a sample is represented by the formula:

$$C = \frac{Q}{\Delta T} \tag{1}$$

where Q is the heat supplied to the samples, and ΔT is the temperature variation during the heating process. It is possible to grasp the difference in heat capacity from the temperature variation in the samples. The measured heat capacities were the orders of SDS, SDBS, DB, and CNC, as seen in Figure 9. The fact that cellulose has a higher temperature increase rate than other surfactants shows that the heat transfer rate is more excellent than that of the others. It is concluded that CNC can be a good option as a dispersant for nanofluid because of its superior properties to other surfactants.

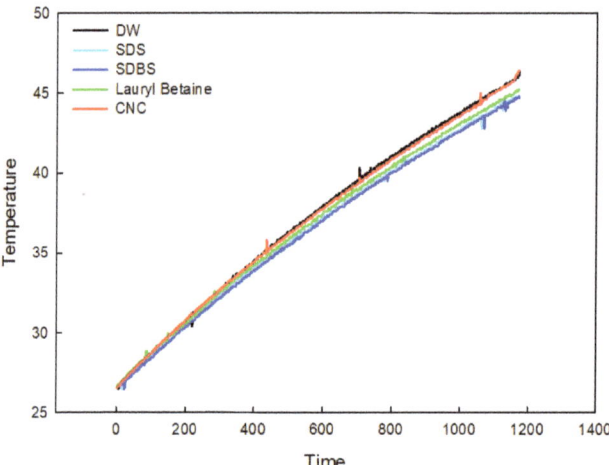

Figure 9. The comparison of heat capacity of various surfactants.

4. Conclusions

The importance of dispersion and stability in the application of nanofluid in industries is becoming more prominent. Accordingly, this study investigated an approach to the nanofluid manufacturing process.

Among these methods, effective dispersion ability was shown by intensively utilizing non-covalent methods rather than strong chemical methods. In this paper, cellulose and surfactants were generally used as non-covalent dispersion methods.

(1) The structure analysis of nano callouses was conducted by the TEM method. It showed rod-shaped nanoparticles acquired from the acid hydrolysis of callouses. The size of the CNC was about 100 nm in diameter and 1000 nm in length. It is shown in Figure 1 that CNC can interact with water strongly by hydrogen bonding because of the hydroxyl group on the molecule.

(2) The electrical conductivity of solutions was studied to figure out the electrical interference in the application of base fluid in industry. CNC had the lowest value of electrical conductivity compared with other surfactants. Furthermore, it was found that, unlike other surfactants, the electrical conductivity of CNC did not change with temperature.

(3) The absorbance of samples was investigated using a UV/VIS spectrophotometer. When other surfactants were used for dispersion, it was hard to use the light absorbance in the ultraviolet range for the evaluation criteria of dispersion and concentration. However, it was revealed that CNC has a stable value across different wavelengths, which indicates that CNC could provide more detailed information on the dispersion and concentration.

(4) The thermal conductivities were examined by the LAMBDA system. First, it was found that the low concentration had high thermal conductivity by comparing the thermal conductivity according to the CNC concentration. As the cellulose content increased, the thermal conductivity of the CNC nanofluid decreased. Overall, CNC had the highest thermal conductivity, followed by LB, and SDS and SDBS had similar thermal conductivity. Moreover, the heat capacity also had a similar value to thermal conductivity, as shown by acquiring the data on differences in temperature when the same quantity of heat was applied.

Therefore, in this study, it was found that cellulose is not only hydrophilic and eco-friendly, but it is also more thermally efficient than other surfactants. These data could be widely used for the base fluid for making nanofluids and in other industries.

Funding: This work was supported by a National Research Foundation of Korea (NRF) grant funded by the Korea government (MSIT) (No. 2022R1C1C2002914).

Conflicts of Interest: The author declares no conflict of interest.

References

1. Khanafer, K.; Vafai, K.; Lightstone, M. Buoyancy-driven heat transfer enhancement in a two-dimensional enclosure utilizing nanofluids. *Int. J. Heat Mass Transf.* **2003**, *46*, 3639–3653. [CrossRef]
2. Zhang, Y.; Zhang, Z.; Yang, J.; Yue, Y.; Zhang, H. A Review of Recent Advances in Superhydrophobic Surfaces and Their Applications in Drag Reduction and Heat Transfer. *Nanomaterials* **2021**, *12*, 44. [CrossRef] [PubMed]
3. Ali, N.; Bahman, A.M.; Aljuwayhel, N.F.; Ebrahim, S.A.; Mukherjee, S.; Alsayegh, A. Carbon-Based Nanofluids and Their Advances towards Heat Transfer Applications—A Review. *Nanomaterials* **2021**, *11*, 1628. [CrossRef] [PubMed]
4. Kim, S.; Song, H.; Yu, K.; Tserengombo, B.; Choi, S.-H.; Chung, H.; Kim, J.; Jeong, H. Comparison of CFD simulations to experiment for heat transfer characteristics with aqueous Al2O3 nanofluid in heat exchanger tube. *Int. Commun. Heat Mass Transf.* **2018**, *95*, 123–131. [CrossRef]
5. Kim, S.; Tserengombo, B.; Choi, S.-H.; Noh, J.; Huh, S.; Choi, B.; Jeong, H. Experimental investigation of heat transfer coefficient with Al_2O_3 nanofluid in small diameter tubes. *Appl. Therm. Eng.* **2019**, *146*, 346–355. [CrossRef]
6. Brandão, A.T.; Rosoiu, S.; Costa, R.; Silva, A.F.; Anicai, L.; Enachescu, M.; Pereira, C.M. Characterization of Carbon Nanomaterials Dispersions: Can Metal Decoration of MWCNTs Improve Their Physicochemical Properties? *Nanomaterials* **2021**, *12*, 99. [CrossRef]
7. Subrahmanyam, K.S.; Ghosh, A.; Gomathi, A.; Govindaraj, A.; Rao, C.N.R. Covalent and Noncovalent Functionalization and Solubilization of Graphene. *Nanosci. Nanotechnol. Lett.* **2009**, *1*, 28–31. [CrossRef]

8. Hu, C.-Y.; Xu, Y.-J.; Duo, S.-W.; Zhang, R.-F.; Li, M.-S. Non-Covalent Functionalization of Carbon Nanotubes with Surfactants and Polymers. *J. Chin. Chem. Soc.* **2009**, *56*, 234–239. [CrossRef]
9. Vázquez-Velázquez, A.R.; Velasco-Soto, M.A.; Pérez-García, S.A.; Licea-Jiménez, L. Functionalization Effect on Polymer Nanocomposite Coatings Based on TiO_2–SiO_2 Nanoparticles with Superhydrophilic Properties. *Nanomaterials* **2018**, *8*, 369. [CrossRef]
10. Rastogi, R.; Kaushal, R.; Tripathi, S.; Sharma, A.L.; Kaur, I.; Bharadwaj, L.M. Comparative study of carbon nanotube dispersion using surfactants. *J. Colloid Interface Sci.* **2008**, *328*, 421–428. [CrossRef]
11. Kalia, S.; Dufresne, A.; Cherian, B.M.; Kaith, B.S.; Avérous, L.; Njuguna, J.; Nassiopoulos, E. Cellulose-Based Bio- and Nanocomposites: A Review. *Int. J. Polym. Sci.* **2011**, *2011*, 837875. [CrossRef]
12. Lin, N.; Huang, J.; Dufresne, A. Preparation, properties and applications of polysaccharide nanocrystals in advanced functional nanomaterials: A review. *Nanoscale* **2012**, *4*, 3274–3294. [CrossRef] [PubMed]
13. Wardhono, E.; Pinem, M.; Kustiningsih, I.; Agustina, S.; Oudet, F.; Lefebvre, C.; Guénin, E. Cellulose Nanocrystals to Improve Stability and Functional Properties of Emulsified Film Based on Chitosan Nanoparticles and Beeswax. *Nanomaterials* **2019**, *9*, 1707. [CrossRef] [PubMed]
14. Siqueira, G.; Bras, J.; Dufresne, A. Cellulosic Bionanocomposites: A Review of Preparation, Properties and Applications. *Polymers* **2010**, *2*, 728–765. [CrossRef]
15. Kim, S.; Tserengombo, B.; Choi, S.-H.; Noh, J.; Huh, S.; Choi, B.; Jeong, H. Experimental investigation of dispersion characteristics and thermal conductivity of various surfactants on carbon based nanomaterial. *Int. Commun. Heat Mass Transf.* **2018**, *91*, 95–102. [CrossRef]
16. Mateos, R.; Vera, S.; Valiente, M.; Díez-Pascual, A.M.; Andrés, M.P.S. Comparison of Anionic, Cationic and Nonionic Surfactants as Dispersing Agents for Graphene Based on the Fluorescence of Riboflavin. *Nanomaterials* **2017**, *7*, 403. [CrossRef]
17. Seelenmeyer, S.; Ballauff, M. Analysis of Surfactants Adsorbed onto the Surface of Latex Particles by Small-Angle X-ray Scattering. *Langmuir* **2000**, *16*, 4094–4099. [CrossRef]
18. Clarke, J.G.; Wicks, S.R.; Farr, S.J. Surfactant mediated effects in pressurized metered dose inhalers formulated as suspensions. I. Drug/surfactant interactions in a model propellant system. *Int. J. Pharm.* **1993**, *93*, 221–231. [CrossRef]
19. Guardia, L.; Fernández-Merino, M.; Paredes, J.I.; Fernández, P.S.; Villar-Rodil, S.; Martinez-Alonso, A.; Tascon, J.M.D. High-throughput production of pristine graphene in an aqueous dispersion assisted by non-ionic surfactants. *Carbon* **2011**, *49*, 1653–1662. [CrossRef]
20. Singh, B.P.; Menchavez, R.; Takai, C.; Fuji, M.; Takahashi, M. Stability of dispersions of colloidal alumina particles in aqueous suspensions. *J. Colloid Interface Sci.* **2005**, *291*, 181–186. [CrossRef]
21. Cushing, B.L.; Kolesnichenko, V.L.; O'Connor, C.J. Recent Advances in the Liquid-Phase Syntheses of Inorganic Nanoparticles. *Chem. Rev.* **2004**, *104*, 3893–3946. [CrossRef] [PubMed]
22. Islam, M.F.; Rojas, E.; Bergey, D.M.; Johnson, A.T.; Yodh, A.G. High Weight Fraction Surfactant Solubilization of Single-Wall Carbon Nanotubes in Water. *Nano Lett.* **2003**, *3*, 269–273. [CrossRef]
23. Yu, J.; Grossiord, N.; Koning, C.E.; Loos, J. Controlling the dispersion of multi-wall carbon nanotubes in aqueous surfactant solution. *Carbon* **2007**, *45*, 618–623. [CrossRef]
24. Li, X.; Zhu, D.; Wang, X.; Wang, N.; Gao, J.; Li, H. Thermal conductivity enhancement dependent pH and chemical surfactant for Cu-H_2O nanofluids. *Thermochim. Acta* **2008**, *469*, 98–103. [CrossRef]
25. Kim, J.-K.; Jung, J.Y.; Kang, Y.T. Absorption performance enhancement by nano-particles and chemical surfactants in binary nanofluids. *Int. J. Refrig.* **2007**, *30*, 50–57. [CrossRef]
26. Peng, H.; Ding, G.; Hu, H. Effect of surfactant additives on nucleate pool boiling heat transfer of refrigerant-based nanofluid. *Exp. Therm. Fluid Sci.* **2011**, *35*, 960–970. [CrossRef]
27. Wang, F.; Han, L.; Zhang, Z.; Fang, X.; Shi, J.; Ma, W. Surfactant-free ionic liquid-based nanofluids with remarkable thermal conductivity enhancement at very low loading of graphene. *Nanoscale Res. Lett.* **2012**, *7*, 314. [CrossRef]
28. Munkhbayar, B.; Tanshen, R.; Jeoun, J.; Chung, H.; Jeong, H. Surfactant-free dispersion of silver nanoparticles into MWCNT-aqueous nanofluids prepared by one-step technique and their thermal characteristics. *Ceram. Int.* **2013**, *39*, 6415–6425. [CrossRef]
29. Yurekli, K.; Mitchell, A.C.A.; Krishnamoorti, R. Small-Angle Neutron Scattering from Surfactant-Assisted Aqueous Dispersions of Carbon Nanotubes. *J. Am. Chem. Soc.* **2004**, *126*, 9902–9903. [CrossRef]
30. Hertel, T.; Hagen, A.; Talalaev, V.; Arnold, K.; Hennrich, F.; Kappes, M.; Flahaut, E. Spectroscopy of Single- and Double-Wall Carbon Nanotubes in Different Environments. *Nano Lett.* **2005**, *5*, 511–514. [CrossRef]
31. Ilyas, R.; Sapuan, S.; Ishak, M.; Zainudin, E. Sugar Palm Nanofibrillated Cellulose Fibre Reinforced Sugar Palm Starch Nanocomposite. Part 1: Morphological, Mechanical and Physical Properties. 2018. Available online: https://europepmc.org/article/ppr/ppr55333 (accessed on 26 April 2022).
32. Ilyas, R.A.; Sapuan, S.M.; Ishak, M.R. Isolation and characterization of nanocrystalline cellulose from sugar palm fibres (Arenga Pinnata). *Carbohydr. Polym.* **2018**, *181*, 1038–1051. [CrossRef]
33. Ilyas, R.A.; Sapuan, S.M.; Ishak, M.R.; Zainudin, E.S. Sugar palm nanofibrillated cellulose (*Arenga pinnata* (Wurmb.) Merr): Effect of cycles on their yield, physic-chemical, morphological and thermal behavior. *Int. J. Biol. Macromol.* **2019**, *123*, 379–388. [CrossRef] [PubMed]
34. Sway, K.; Hovey, J.K.; Tremaine, P.R. Apparent molar heat capacities and volumes of alkylbenzenesulfonate salts in water: Substituent group additivity. *Can. J. Chem.* **1986**, *64*, 394–398. [CrossRef]

35. Bakshi, M.S.; Crisantino, R.; De Lisi, R.; Milioto, S. Volume and heat capacity of sodium dodecyl sulfate-dodecyldimethylamine oxide mixed micelles. *J. Phys. Chem.* **1993**, *97*, 6914–6919. [CrossRef]
36. Garg, P.; Alvarado, J.L.; Marsh, C.; Carlson, T.A.; Kessler, D.A.; Annamalai, K. An experimental study on the effect of ultrasonication on viscosity and heat transfer performance of multi-wall carbon nanotube-based aqueous nanofluids. *Int. J. Heat Mass Transf.* **2009**, *52*, 5090–5101. [CrossRef]
37. Tserengombo, B.; Jeong, H.; Dolgor, E.; Delgado, A.; Kim, S. Effects of Functionalization in Different Conditions and Ball Milling on the Dispersion and Thermal and Electrical Conductivity of MWCNTs in Aqueous Solution. *Nanomaterials* **2021**, *11*, 1323. [CrossRef]
38. Molnes, S.N.; Paso, K.G.; Strand, S.; Syverud, K. The effects of pH, time and temperature on the stability and viscosity of cellulose nanocrystal (CNC) dispersions: Implications for use in enhanced oil recovery. *Cellulose* **2017**, *24*, 4479–4491. [CrossRef]
39. Bentley, J.P. Temperature sensor characteristics and measurement system design. *J. Phys. E Sci. Instrum.* **1984**, *17*, 430. [CrossRef]

MDPI
St. Alban-Anlage 66
4052 Basel
Switzerland
www.mdpi.com

Nanomaterials Editorial Office
E-mail: nanomaterials@mdpi.com
www.mdpi.com/journal/nanomaterials

Disclaimer/Publisher's Note: The statements, opinions and data contained in all publications are solely those of the individual author(s) and contributor(s) and not of MDPI and/or the editor(s). MDPI and/or the editor(s) disclaim responsibility for any injury to people or property resulting from any ideas, methods, instructions or products referred to in the content.

www.ingramcontent.com/pod-product-compliance
Lightning Source LLC
LaVergne TN
LVHW070558100526
838202LV00012B/499